From Discovery to Cure: The Early Days of Physics in Radiology

Parkar

Copyright © 2024 by Parkar

Contents

Chapter 1

Introduction

Physics applied to radiology begins with the discovery of *X-rays* by Wilhelm Röntgen (1885), radioactivity by Henri Becquerel (1886), and the ^{226}Ra isotope by Marie and Pierre Curie (1889). These discoveries were quickly developed and implemented in the practice of medicine.

Years later, in 1896, the first external radiotherapy occurred, where *X-rays* were used to treat breast cancer. The first treatments were carried out with sources of low penetrating power. With the development of *X-ray* tubes (with voltages between 180 and $200kV$), where Coolidge was a pioneer, the penetration power increased. It was possible to give more energy to deeper tumors. Later, and after a transition by cobalt devices ($1.3MV$), linear accelerators (with voltages between 4 and $20MV$) were introduced.

The additional development of accelerators drove new advances and techniques such as conformal radiation therapy (CRT), intensity-modulated radiation therapy (IMRT), $4D$ radiation therapy, and others, contributing to greater use of this modality [1].

Recently, ion beam radiotherapy, such as proton therapy and carbon therapy, has revealed fascinating characteristics, and it is starting to be implemented more often in clinical spaces.

1.1 Context

The main argument for using particles (protons/ions) in radiotherapy is their ability to lower the deposited dose to surrounding healthy organs. However, due to the steep dose gradient at the distal edge of *Bragg* peak, uncertainties in the particle range's determination can have a profound impact on healthy surrounding organs [2–4].

A situation without and with the influence of uncertainties is represented in figure 1.1. The figure compares the dose deposition profile in depth for photons and protons. Uncertainties are a big issue for protons when compared with photons. On the other hand, in a situation without uncertainties, the potential of tissue spare is significantly higher for protons. To cover the entire tumor, treatments with protons use different *Bragg* peaks corresponding to different depths (energies). The combined total dose of all *Bragg* peaks is called the Spread-Out *Bragg* Peak (SOBP). Figure 1.1 shows that SOBP has the same advantages and disadvantages of a single energy *Bragg* peak [5].

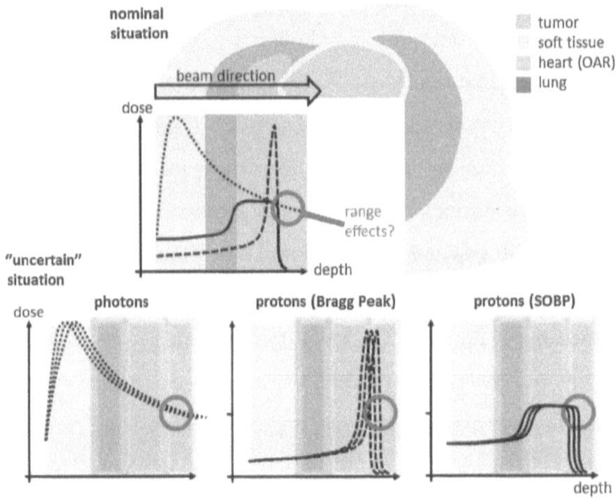

FIGURE 1.1: Proton treatment compared with photon treatment without (a) and with (b) influence of uncertainties: dotted line: photon depth dose curve; dashed line, mono-energetic proton depth dose curve (*Bragg* peak); straight line: spread out proton *Bragg* curve (SOBP). Adapted from [5].

In a recent paper, the inter-center variation of range errors was compared with a ground-truth phantom scanned at 17 participating proton centers. Alarmingly, the 2σ inter-center variation in range prediction was 2.6% and 2.9% for typical brain and prostate treatment fields [6].

The Bragg peak uncertainty problem is a challenge many research groups have been engaged in. There are two significant lines of research in range monitoring technology anatomic-tracking methods and beam-tracking methods.

Anatomic-tracking methods produce $3D$ images of the patient's particle effect/absorption via imaging technologies, such as positron emission tomography, proton/ion computer tomography, or magnetic resonance imaging. From which they infer the particle range [7–9]. Regardless, these methods are not immediate and suffer from biological washout and organ motion within the patient.

Beam-tracking methods are primarily based on the prompt-gamma (PG) rays emitted when particles interact with a medium via nuclear reactions. In these methods, no image is produced, and it is required a localization system to allow the merging of the beam path with the patient $3D$ anatomy.

1.2 Motivation and Objectives

Recently, prompt gamma-ray imaging (PGI) has emerged as a promising method for *in-vivo* range monitoring [5]. The beam-tracking method PGI has been proposed as a superior method to the current anatomic-tracking methods [10]. This relies on the fact that the nuclear reaction positions are correlated with the deposited dose [11]. Thus, numerous induced prompt-gamma ray emissions within the nano-second scale can be detected to infer the dose (or range).

Diverse solutions have been suggested for PGI [10]. Prompt gamma-ray spectroscopy (PGS) is among the most promising solutions for *in-vivo* range monitoring. With some maturity, PGS has been highlighted as a technique that successfully measured absolute range deviations [11–14].

Meanwhile, the strong presence of range uncertainties in prostate cancer treatments hinders the exploration of new techniques in particle therapy of prostate cancer (e.g., use of anterior beams instead of the commons bilateral beams). Interplay effects play a major role in range uncertainties associated with prostate cancer treatments. Thus, a rectal balloon filled with water is often used. When inserted in the rectum, the endo-rectal balloon can fix the prostate position and lower the anatomic variations between and during fractions [15].

The re-use of the endo-rectal balloon together with PGS can be exploited to enable range control. By replacing the water in the rectal balloon with a silicon solution, it is possible to use PGS to pickup the silicon signal. The silicon signal is visible for relative lower energies, where the strongest PG line lays on 1.78 MeV. During treatment, if the silicon spectral signal is measured that means that *Bragg* peak has crossed the rectal wall into the rectal balloon. Thus, it is necessary to reduce the proton beam energy to pull back the Bragg peak until there is no longer a silicon signal in the spectrum. With this technique, the PGS signal can be used to control the Bragg peak location and reduce the rectal wall dose. Along these lines, the main goal of this book is the first study of an *in-vitro* setup entailed with proton beams, PGS measurements and the use of a rectal balloon filled with a ^{28}Si-based solution within a prostate-like phantom. To achieve such goal the ^{28}Si line in PG spectra will be studied in qualitative and quantitative manner.

This book is organized as follows. Chapter 2 (*Background: Medical Physics*) gives a general background in physical principles related (directly and indirectly) to particle therapy, or more specifically, proton therapy. Emphasis is given to subjects underline in this book. Chapter 3 (*Prompt Gamma-Ray Imaging: State of the Art*) presents the state of the art of prompt gamma-ray imaging (PGI). Chapter 4 (*Proton Therapy in Prostate Cancer*) follows a more in-depth understanding of proton therapy for prostate cancer. The methods and materials used and developed in this book are present in Chapter 5 (*Materials and Methods*). Chapter 6 (*Results and Discussion*) shows the *in-vitro* results of PGS experiments in prostate cancer. Furthermore, Chapter 6, evaluates and discusses all results associated to the ^{28}Si PG line emission. Finally, Chapter 7 (*Conclusion and Future Work*) presents the future work to be developed and the various book conclusions.

Chapter 2

Background: Medical Physics

Presented in this section are the main medical physics concepts underlying this book Since this book is centered on particle therapy, so is this section, which is intended to provide basic knowledge about the physics that governs its aspects. Further-more, this section will address diverse medical physics subjects for future references.

2.1 Interaction of Radiation with Matter

When x or γ radiation travel through matter, their energy can be transferred to the medium as a result of physical interactions of the radiation with it. Of medical physics interest, there are three fundamental interaction mechanisms of the radiation with matter, the photoelectric effect, the *Compton* effect, and the pair production. The probability of these interactions happening can be expressed as a cross-section, and it depends on the photon energy and electronic density of the medium.

2.1.1 Photoelectric Effect

The photoelectric effect is characterized by the total energy transference of a photon to a bound electron. As a result, the electron is ejected (with energy equal to the difference between the photon and bound energy), which causes the atom to become ionized. Following this interaction is the emission of a photon with an energy matching up the difference between the unsaturated level and the electron energy at the upper level. The last will occupy the vacancy created by the photoelectron, and a repeated process will occur until the whole atom reaches an energy level equivalent to its lowest bounded energy. This process can release further electrons from the atom. Known as the *Auger* effect, the release of chain electrons is the most probable effect for elements with lower atomic numbers. The probability (cross-section), σ, that a photoelectric effect will occur depends upon photon energy, E_λ, and atomic number, Z (see equation 2.1). Usually, and due to different effectiveness within different energies, n and m in equation 2.1 are within the range of 3 to 5 depending upon energy ($n = 4$ and $m = 3$ for low energies, $n = 5$ and $m = 1$ for high energies) [16].

$$\sigma \propto Z^n / E_\lambda^m \tag{2.1}$$

2.1.2 *Compton* Effect

One can consider an atom electron approximately free when colliding with an incident photon if the photon energy, E_x, is significantly higher than the electron binding energy. The laws of conservation of momentum and energy are valid in such a process. Therefore, and knowing that the photon is massless, its total absorption is not possible. As a result, the photon has to be dispersed, losing some of its energy. Termed by the *Compton* effect (see figure 2.1), the before-mentioned interaction most likely occurs in the electrons of the outermost shells.

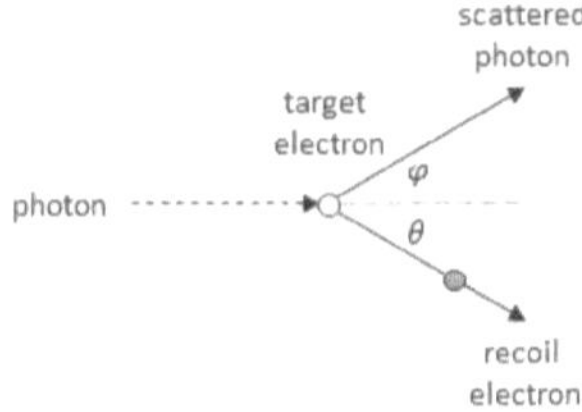

FIGURE 2.1: Schematic illustration of a *Compton* interaction.

The incident photon (with linear moment, $p_x = E_x/c$) upon reaching the stationary electron causes its ejection with energy E_e and angle θ. As a result, the photon is scattered with an angle φ, energy E_x' and liner momentum p_x'. Mathematics relations regarding the *Compton* interaction can be expressed by equations 2.2, 2.3 and 2.4 where $m_e c^2$ is the electron rest energy, and units are in MeV.

$$E_x' = \frac{E_x}{1 + \frac{E_x}{m_e c^2}\left(1 - \cos \varphi\right)} \tag{2.2}$$

$$E_e = \frac{E_x}{\frac{m_e c^2}{E_x} + 1 - \cos \varphi} \left(1 - \cos \varphi\right) \tag{2.3}$$

$$\cos \theta = \left(1 + \frac{E_x}{m_e c^2}\right) \frac{\tan \theta}{2} \tag{2.4}$$

In a first approach, Thomson's description was used to obtain the differential cross-section of the Compton effect. However, this cross-section was an approximation and had significant errors for angles, $\theta \neq 0$ [17]. Later, in 1928, Klein and Nishina added an atomic form factor, F_{KN}, leading to a better theory expressed by equation 2.5, where r_e is the classical radius of the electron and $\frac{d_e \sigma_C^{KN}}{d\Omega}$ is the electronic differential cross-section per unit solid angle for *Compton* effect [17].

$$\frac{d_e \sigma_C^{KN}}{d\Omega} = \frac{r_e^2}{2} \left(1 + \cos^2 \theta\right) F_{KN} \tag{2.5}$$

Figure 2.2 shows a representation of a radiation detector response to a monoenergetic beam of γ-rays. Regarding the *Compton* effect, one can see a *Compton* continuum due to all scattering angles that occur on the detector. Furthermore, the edge before the full energy peak represents a head-on collision in which $\theta = \pi$ (γ-ray is backscattered) heading to a maximum energy transference predicted by the *Compton* effect.

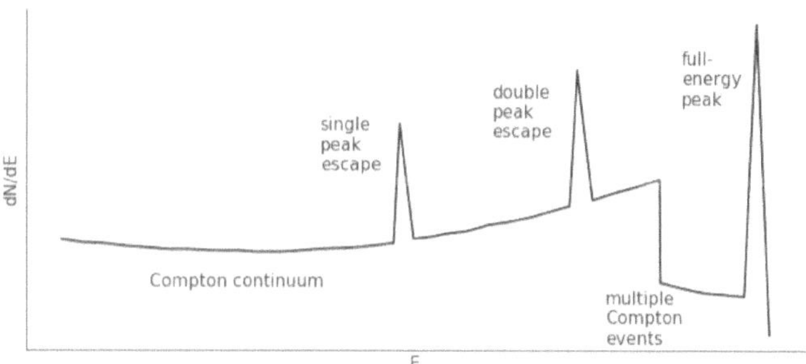

FIGURE 2.2: Schematic representation of a typical radiation detector response to a monoenergetic beam of γ-rays. Adapted from [18].

2.1.3 Pair Production

When an incident photon exceeds the threshold energy of $2m_0 c^2 = 1.022\,MeV$, the creation of a particle-antiparticle pair (e.g., positron-electron) with the total absorption of the photon becomes energetically possible. To make such processes possible, the laws of energy conservation, charge, and moment are valid. As a consequence of the previous stipulations, this type of interaction arises only in the scattering centers' *Coulomb* field. If the interaction occurs with the target's electron field instead of the nuclear one, it is called triplet production. Afterward, the pair production, the antiparticle, and particle will lose their kinetic energy through interactions with the medium. At this point, the antiparticle can be considered at rest and consequently is annihilated. That is, the cross-section for positron annihilation has a maximum for the situation where the positron is at rest. This process, depending on the intrinsic angular momentum of the particles, will result in an odd or even number of photons (usually two photons) traveling in opposite directions.

Pair production has a vital role in detection since it affects it so that noise arises due to annihilation occurring in the detector. If one or two photons resulting from annihilation escape the detector without any interaction, respectively, a single (SE) or a double escape peak (DE) will rise (see figure 2.2). In an ideal case, one wants a complete absorption of the photons.

2.2 Interactions of Charged Particles with Matter

When traveling through the matter, charged particles interact with targets (atomic orbital electrons, atomic nuclei, or atoms as a whole) and, as a result, lose energy and suffer multiple deflections. These particles (or projectiles) in the medical physics context are light charged particles (*e.g.*, electrons, positrons) and heavy charged particles (*e.g.*, protons, carbon ions). The interactions between the charged particles and targets can be classified as nuclear reactions and elastic or inelastic collisions. In general, light and heavy charged particles interact with atomic nuclei and atomic orbital electrons through *Coulomb* interactions. These interactions in classical physics can be seen as a collision between a projectile (charged particle) and a stationary target. The collisions are correlated with a probability of occurrence given by a cross-section that is a function of both charged particles and target physical properties.

2.2.1 *Coulomb* Interactions of Charged Particles

The dominant interactions in charged particles are elastic and inelastic collisions mediated by *Coulomb* forces. In terms of dosimetric manifestations, elastic interactions determine lateral penumbral sharpness, while inelastic interactions are associated with energy loss and determine the particle range in the patient. Regarding the energy transfer from charged particles to a medium is useful to divide the *Coulomb* interactions into three categories (shown schematically in figure 2.3) depending on the relative size of the classical impact parameter b when compared with the atomic radius a (see figure 2.3) [17, 19].

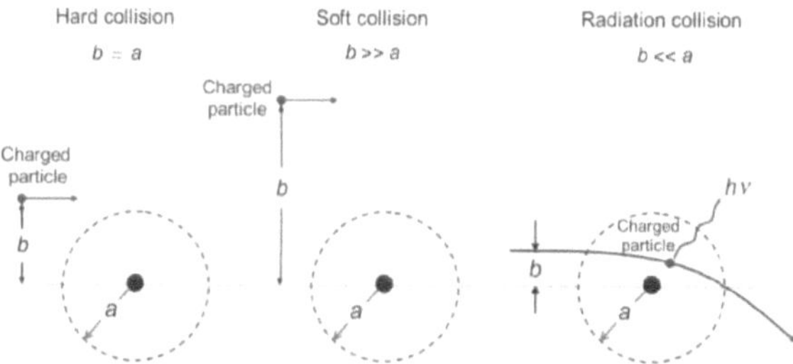

FIGURE 2.3: Schematic illustration of charged particles interactions with matter: hard collision ($b \approx a$); soft collision ($b \gg a$); radiation collision ($b \ll a$). Adapted from [17].

2.2.1.1 Soft/Distant collision ($b >> a$)

The charged particle is far from the atomic nuclei, and the interaction occurs with the atom as a whole. Even though the energy loss is small in this type of interaction, the number of interactions is large. It is one of the most common interactions when a charged particle travels through a medium. The results of these interactions are either atomic polarization, excitation, or ionization by ejecting a valence-shell electron [17, 19].

2.2.1.2 Hard/Close collision ($b \approx a$)

The charged particle has a high probability of interacting with orbital electrons, transferring a significant amount of energy. The eject electrons that result from this interaction are commonly referenced as δ-rays. Unlike soft collisions, hard collisions do not happen so often, but on the other hand, the energy transfer associated with it is higher [17, 19].

2.2.1.3 *Bremsstrahlung* **production or radiation collision** ($b << a$)

The charged particle interacts with the nucleus and suffers either elastic or inelastic scattering. Contrarily to heavy charged particles, where this interaction is negligible, in light charged particles, *Bremsstrahlung* production is relevant and needs to be considered ($probability \propto mass_{chargedparticle}^{-1}$). Most of the radiation interactions are elastic interactions where the scattered particle loses only an insignificant amount of kinetic energy (also known as **multiple scattering of charged particles**) [20]. Nevertheless, some of the radiation interactions are inelastic, followed by an *x-ray* photon. In this last case, also known as *Bremsstrahlung* collision (break), the energy loss can be significant [17, 19].

2.2.2 Nuclear Reactions

Nuclear reactions are governed by conservation of energy and momentum laws. Physical quantities like charge, linear momentum, and mass-energy are conserved. Most important are electric, baryonic and leptonic charges. Nuclear reactions can be characterized by a collision between a particle and a target, resulting in excitation of nuclide (induced gamma emission) or transformation of at least one nuclide to another (spallation). The nuclear reaction energy or Q value is defined by subtracting the rest energies of products after the reaction from the rest energies of products before the reaction (see equation 2.6, where m_i is the mass of the i particle and c the light speed constant) [17, 19].

The Q value (nuclear reaction energy) is a parameter that defines the type of collision. This is: for $Q > 0$, the collision is exothermic (release of energy); for $Q < 0$, the collision is endothermic (requires a transference of energy from the projectile to the target); and for $Q = 0$ the collision is elastic. Differently from exothermic reactions, where the reaction can occur spontaneously, the endothermic reactions only happen when the kinetic energy of the projectile is greater than certain threshold energy, $E_{k,thr}$, given by equation 2.7, where m_p and m_t are the mass of the particle (or projectile) and target, respectively [17, 19].

$$Q = \sum_{i,before} m_i c^2 - \sum_{i,after} m_i c^2 \tag{2.6}$$

$$E_{k,thr} \sim -Q \left(1 + \frac{m_p}{m_t} \right) \tag{2.7}$$

From a therapeutic perspective, nuclear reactions have a small effect on the absorbed dose and are not so important as the interactions mentioned above. They take place before the end-of-range and result in a slight decrease of the absorbed dose (or beam fluence) due to the removal of primary protons. In order for a nuclear reaction occur, the induced particle needs to overcome the *Coulomb* barrier of the targets and the total non-elastic cross-section threshold energy of the endothermic reaction (see equation 2.7). The total cross-section, $\sigma_N(E)$, that governs nuclear reactions can be factorized to a simple energy dependence according to equation 2.8, where E is the particle energy, η, the *Sommerfeld* parameter and $S(E)$ the astrophysical S-factor [13, 21].

$$\sigma(E) = \frac{1}{E} e^{-2\pi\eta} S(E) \tag{2.8}$$

2.2.2.1 Prompt Gamma-Ray by Particle Induced Reactions

A sub-product of nuclear reaction is prompt gamma-ray (PG) emission. Knowing the effective cross-section for the emission of PG, $\sigma_\gamma(E)$, the production of PG, dG per unit distance, dx is given by equation 2.9, where Z_{eff} is the effective atomic number of the target and u is the atomic mass unit [22]. The cross-section besides the threshold energy (lower than $\sim 10 MeV$ for atomic nuclei relevant for this book), rises rapidly to a maximum and then asymptotically decreases for higher energies [23–25].

$$\frac{dG}{dx} = \sigma_\gamma(E) * \frac{\rho(x_p)}{Z_{eff} u} \tag{2.9}$$

The intense study of the PG emission due to particle-induced reactions has not been the subject of any solo research in particle therapy. Notwithstanding, several groups have researched these reactions and their cross-sections due to a specific interest in astronomy. Among the most studied are the nuclear reactions induced by protons in targets such as ^{12}C, ^{14}N and ^{16}O. Coincidently and jointly with ^1H, the former nuclides are the most abundant in the human body and therefore of particular interest for this book.

2.2.2.2 Individual Gamma-Ray Lines

Individual gamma-ray lines due to nuclear reactions are resumed in Appendix A. The most significant and important lines (regarding this book) in ^{12}C, ^{14}N and ^{16}O are the 0.718, 1.02, 1.38, 1.64, 2.00, 2.31, 4.44, 5.27, and 6.13 MeV lines as given by tables A.1, A.2 and A.3. It is important to note that broadening happens due to different velocities of primary particles line broadening. Therefore, some lines cannot be distinguished from neighboring lines.

Proton-induced ^{28}Si gamma-ray lines are once more of distinct interest for this book A summary of this lines is presented in Appendix A, table A.4. The strongest line produced on ^{28}Si is at 1.78 MeV. Created by an inelastic collision, results from a transi-tion from the first excited state to the ground state. The threshold for this transition is 1.8 MeV. Adding this energy to the *Coulomb* barrier (≈ 2.3 MeV), the cross section reaches significant values above 3 MeV (see figure 2.4) [26].

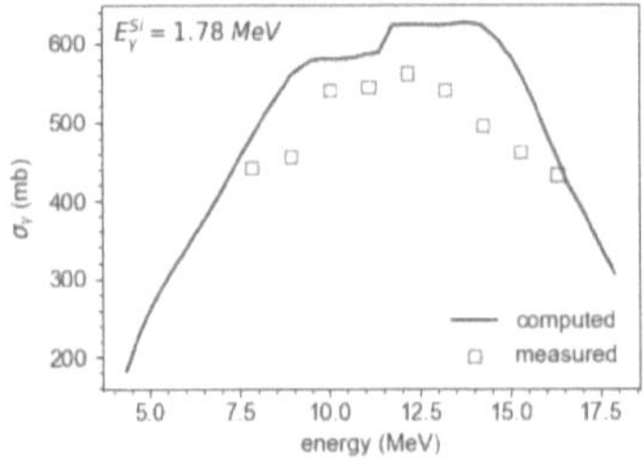

FIGURE 2.4: Proton induced gamma-ray cross-section for 1.78 MeV transition in ^{28}Si: blue line: calculate; red square: measured. Adapted from [26].

2.2.3 Stopping Power

The *Coulomb* interactions of charged particles in an absorbing medium give rise to energy losses. The rate of this energies losses per unit of path length dx in a medium is stated as linear stopping power ($-dE\ /\ dx$). The quotient, linear stopping power per density of the absorbing medium, ρ, results in a quantity called mass stopping power S_{total} with units of $MeV \cdot cm_2 \cdot g^{-1}$. Stopping power is a propriety of the absorbing medium. When associated with charged particles, two types of stopping power are known: radiation stopping power (or nuclear stopping power), S_{rad}, resulting from *Bremsstrahlung* production; and collision stopping power (ionization/electronic stopping power), S_{col}, resulting from *Coulomb* interactions with atomic orbital electrons of the absorbed. Thus, in general, the total stopping power can be expressed by the equation 2.10.

$$S_{total} = \frac{-dE}{dx \cdot \rho} = S_{rad} + S_{col} \tag{2.10}$$

2.2.3.1 Radiation (Nuclear) Stopping Power

Radiation stopping power is related to the probability of *Bremsstrahlung* emission and, therefore, with its cross-section, σ_{rad}. Hans Bethe and Walter Heiltler in 1930s developed a theoretical model for the radiation stopping power (see equation 2.11, where N_a is the atomic density and σ_{rad} the cross-section for *Bremsstrahlung* production) [17, 27]. The contribution of radiation losses, when compared to collision losses, is negligible for heavy charged particles ($S_{rad} \approx 0$), making the radiation stopping power not a subject of interest in this study.

$$S_{rad} = N_a \sigma_{rad} E \tag{2.11}$$

2.2.3.2 Collision (Electronic) Stopping Power

Bragg and Kleeman made the first mathematical approach to the determination of collision stopping power in 1905 [28]. Later on, in 1915, Bohr developed a more physically complete theory based on impact parameter b between the charged particle and the absorber nucleus [29]. Bohr's theory in intermediary energies showed a good agreement with empirical data. However, for lower and higher energies, the theory had a significant discrepancy compared with experimental data [17]. The lack of quantum mechanical and relativistic effects in Bohr's theory was the main reason for this discrepancy. Addressing quantum and relativistic effects, Hans Bethe (1930) and Felix Bloch (1933) come up with a more accurate theory represented by the equation 2.12, where N_A is *Avogadro*'s number, A the atomic weight of the target, ε_0 the vacuum permittivity, I the mean ionization/excitation energy*, and $\beta = v/c$ where v is the projectile velocity. It is essential to mention that equation 2.12 are valid only for heavy charged particles [17, 30, 31].

$$S_{col} = 4\pi \frac{ZN_A}{A} \left(\frac{q_e^2}{4\pi\varepsilon_0} \right)^2 \frac{z^2}{m_e\beta^2c^2} \left[\ln \frac{2m_ec^2}{I} + \ln \frac{\beta^2}{1-\beta^2} - \beta^2 \right] \tag{2.12}$$

---

*The mean ionization/excitation energy, I, is determined empirically from the measurement. Usually, (International Commission on Radiation Units) ICRU and National Institute of Standards and Technology (NIST) tables are used to estimate I. However, it is also possible to use empirical mathematical approximations.

Although equation 2.12 is a good approximation, it is still not in complete agreement with experimental data. Several investigations have been carried out to correct this problem, being the most important the shell correction (C/Z) and the density effect correction (δ). Including the mentioned modifications, one can write equation 2.13 that translates into a better theory for the collision stopping power of heavy charged particles [17].

$$S_{col} = 4\pi \frac{ZN_A}{A} \left(\frac{q_e^2}{4\pi\varepsilon_0}\right)^2 \frac{z^2}{m_e\beta^2c^2} \left[\ln\frac{2m_ec^2}{I} + \ln\frac{\beta^2}{1-\beta^2} - \beta^2 - \frac{C}{Z} - \delta\right] \qquad (2.13)$$

Figure 2.5 exhibit the standard collision stopping power usually described by three regions. At lower energies, S_{total} rises and reaches a peak, in the intermediate energy S_{total} decrease as a function of $1/v^2$, and finally, the last terms of equation 2.12 influence the behaviour of stopping power for relativistic energies.

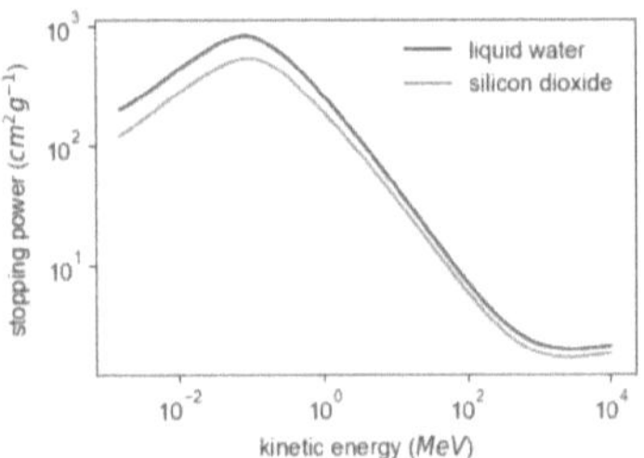

FIGURE 2.5: Stopping power versus energy for protons in liquid water and silicon dioxide. Data were obtained from the NIST database [32].

2.2.4 Bragg Peak

The rate of stopping power caused by ionizing interactions for charged particles is inversely proportional to the square of its speed, β^{-2} (the most important term in equation 2.13). Thus, as the particle slows down, the rate of energy lost increases, as does the ionization of the absorbent medium. In figure 2.6 is represented the dose deposition of a proton and carbon as it penetrates the matter. In its trajectory, the particle slowly deposits its energy, later, near the end-of-range, is produced a peak followed by a distal falloff. This final effect is called the Bragg peak and has a vital role in ion particle treatment. To cover the entire tumor, treatments with protons use different *Bragg* peaks corresponding to different depths (energies). The total dose of all *Bragg* peaks in treatment is called the Spread-Out *Bragg* Peak (SOBP).

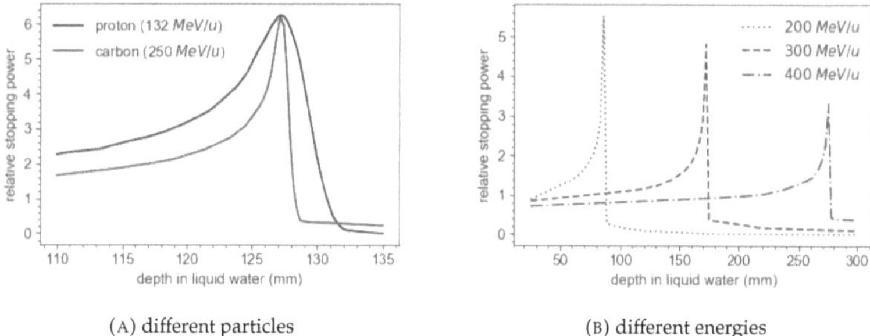

(A) different particles   (B) different energies

FIGURE 2.6: A: *Bragg* peak curve for proton (132 MeV/u) and carbon (250 MeV/u) particle with same range in liquid water. B: *Bragg* peak curve for carbons with different energies. Adapted from [33].

2.2.5 Particle Range

One can define the range (or mean projected range) of a particle in a medium, R, as the total distance traveled by a particle until it comes to rest. On the other hand, when traveling throughout the medium, the charged particle undergoes several collisions, where, between them, it travels small distances losing its energy continuously. The former concept is known as "continuous slowing down approximation" (CSDA). It differs from the range concept in a way wherein only the initial, and the most distal distance is considered (see figure 2.7). The CSDA range, R_{CSDA}, is defined in equation 2.14 [17, 19].

$$R_{CSDA} = \int_0^{E_k^0} S_{tot}^{-1} \cdot dE \tag{2.14}$$

From the CSDA point of view, heavy charged particles experience small deviations in their trajectories. Nevertheless, these deviations are minimal and almost do not change their direction. To a certain extent, it can be said that R_{CSDA} is approximately equal to R. Thus, the integral in equation 2.14 represents a good approximation of the range for these particles. As particles go in a CSDA, a transference of energy per unit distance occurs. In dosimetry, this transference of energy is known as linear energy transfer (LET) and is closely related to the stopping power [17, 19].

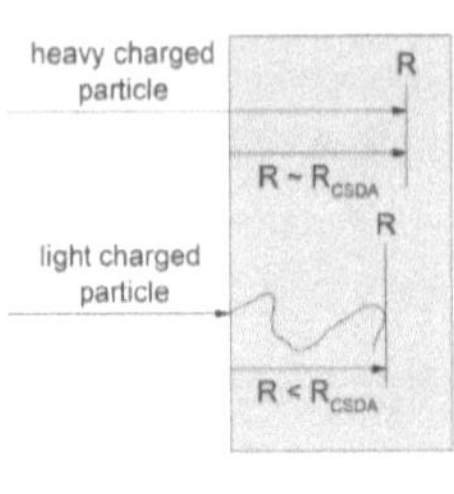
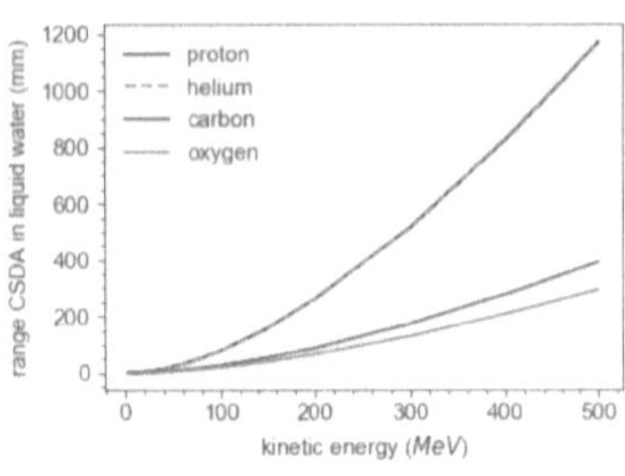

(A) different particles

(B) different energies

FIGURE 2.7: A: schematic diagram of a heavy (e.g.: proton) and a light (e.g.: electron) charged particle travelling throughout the medium. Adapted from [17]. B: CSDA range of protons as function of their kinetic energy in silicon dioxide and liquid water. Data obtained from the NIST database [32].

2.2.5.1 Energy Straggling

Energy straggling (or range straggling) is one of the physical processes that strongly governs the shape of a particle *Bragg* curve. The approximation of a smooth and continuous slowing down may be valid for clinical calculations. However, stochastic fluctuations of the energy loss occur and have an impact on *Bragg* curve. This phenomenon is known as energy straggling and results in slightly different total path lengths for each particle. The energy straggling is dependent on penetration depth and mass of the particle (varies approximately as the inverse of the square root). Thus, higher energies (or ranges) are related with *Bragg* peaks of larger width and smaller height (see figure 2.6). Regarding different ions, higher the mass, sharper is the peak (see figure 2.6) [33, 34].

2.2.5.2 Lateral Scattering

The multiple scattering of charged particles (mention before) spreads an initially parallel beam of heavy charged particles into an approximated *Gaussian* angular distribution [19]. This phenomenon is known as lateral scattering and is dependent on particle energy and mass. In figure 2.8, one can observe that due to their higher mass, heavier charged particles have lower lateral scattering.

Unlike energy straggling, lateral scattering is clinical relevant for treating of tumors close to organs at risk (OAR).

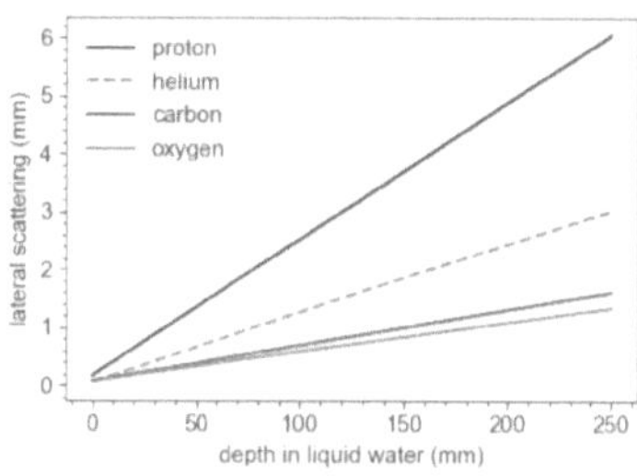

FIGURE 2.8: Lateral scattering in liquid water for different particles. Adapted from [33].

2.2.5.3 Ion Fragmentation

Nuclear reactions are responsible for ion fragmentation. At high energies, spallation reactions may result in the disintegration of the projectile and target nuclei (creation of secondary particles). Regarding protons, only target nuclei can suffer from fragmentation, and therefore, comparing to heavier charged particles, the ion fragmentation of protons is low. It is important to notice that fragmentation has its role in the treatment plan because fragments deposit energy beyond the distal edge of the Bragg peak [35].

2.3 Particle Therapy

Particle therapy, first proposed in the 1940s, is an emerging technique of external beam radiotherapy [36]. It is widely accepted (more than 200000 patients are treated per year in 100 facilities worldwide) and promising for tumor treatments [37].

As seen in figure 2.6, in particle therapy, the dose at the entrance point is minimum and maximum close to the particle range. This effect results from higher ionization cross sections for low particle velocities. The dose deposition falls steeply to zero beyond *Bragg* peak and increases laterally with depth. Particle energy, stopping power, target density, and composition are the main physical properties that define the dose in tissues. Another distinguishing feature of particle therapy is the production of secondary products, such as neutrons, annihilation photons, and prompt gamma-rays (PG) [38].

Different features make particle therapy a desired choice over conventional photon therapy. These are the following [38]:

- *distal edge*:

    - proton therapy: steep dose gradient promising for OARs sparing;

     − photon therapy: slowly falling depth dose curve of photons limits sparing of OARs;

- *integral dose*:

     − proton therapy: dose can be focused on the tumor volume;

     − photon therapy: large volume of tissue exposed to dose from several directions.

Despite the advantages, particle therapy has it owns drawbacks [38]:

- facility and beam operation with high costs;

- lack of large clinical trials and evidence about the superiority of particle therapy;

- ranges uncertainties due to patient, organ motion and tissue composition:

     − stopping power ambiguity;

     − patient alignment errors;

     − anatomy changes between or during fractions;

     − organ motion in the thoracic and abdominal region;

     − biological factors.

2.3.1 Therapeutic Concepts

In the next sub-sections, some therapeutic concepts relevant to this book are succinctly present.

2.3.1.1 Dose

International Commission on Radiation Units (ICRU) defines absorbed dose as absorbed energy, dE per unit mass, dm (see equation 2.15). It has units of $Gy = J/kg$. The absorbed dose is relevant for all fields of ionizing radiation, whether direct or indirect ionization, as well as for any source of ionizing radiation distributed within an absorption medium.

$$D = \frac{dE}{dm} \tag{2.15}$$

2.3.1.2 Relative Biological Effectiveness

Chemical and biological processes caused by radiation depend not only on the absorbed dose but also on its distribution. The efficiency with which ionizing radiation provokes an individual chemical or biological response is known as relative biological effectiveness. RBE is defined as the ratio of X-ray and particle dose that produces the same biological effect (see equation 2.16). When compared with photons, protons and heavier particles have a higher radiobiological effect on tissue.

$$RBE = \frac{D_{X-ray}}{D_{particle}} \tag{2.16}$$

2.3.1.3 Tumor Control Probability and Normal Tissue Complication Probability

Tumor control probability (TCP) measures the likelihood that a specific radiation dose will provide tumor control or eradication. Separately, normal tissue complication probability (NTCP) measures the likelihood of adverse side effects on healthy tissue for a specific dose. In figure 2.9 is represented the curves of TCP and NTCP in the function of the dose.

2.3.1.4 Dose Volume Histograms

Dose-volume histograms (DVH) are a graphical representation of the planning target volume (PTV) or tissue percentage that receives a given dose. In figure 2.9 is represented as a typical DVH for prostate cancer with proton therapy. It is usually to define D_a and V_b as absorbed dose by $a\%$ of volume and volume receiving $b\%$ of the dose, respectively.

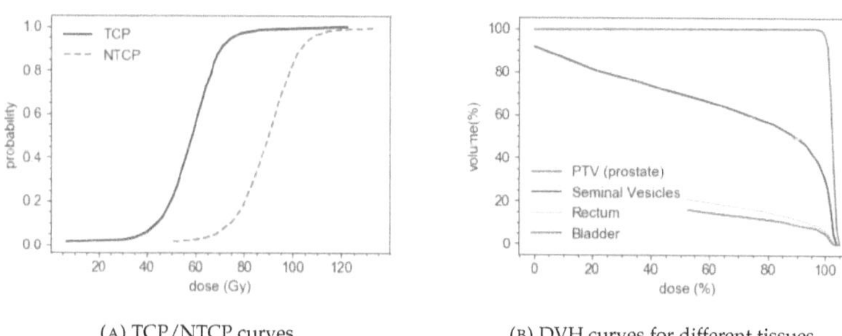

(A) TCP/NTCP curves

(B) DVH curves for different tissues

FIGURE 2.9: A: Sigmoidal shaped response curves for TCP and NTCP. B: Example of DVH for prostate (PTV), rectum and bladder in proton therapy (lateral beams). Adapted from [39].

2.3.1.5 Particle Accelerators

The particle beam is produced by particle accelerators such as cyclotrons, synchrotrons or less common synchrocyclotrons. Once in the beamline, the beam is conformed, collimated and guided to the treatment room. There are usually four or five treatment rooms. They are equipped with fixed horizontal nozzles and/or g antrys. The gantry is an electromechanical system (e.g., arm) that allows a beam incident with a total degree of freedom of 360°.

A cyclotron consists of two dipoles magnets in a 'D' shape and faced to each other by their straight sides. The dipoles produce a uniform magnetic field. Throughout an oscillating electric field, the particles are accelerated in a semicircular p ath. The particle path gets broader as the energy increases (or acceleration), but its oscillating time transition is kept during the process. The particle's energy is fixed, and it is needed an additional system to achieve different energies (up to $\approx 250\,\mathrm{MeV/u}$). The energy selection system uses a degrade of variable thickness to set the final beam e nergy. The beam is delivered collimated, and in short bunches, lasting nanoseconds [40, 41].

A synchrotron, contrary to cyclotrons, has a continuous energy level selection. It consists of a circular ring where electromagnetic resonant cavities accelerate the particles. The beam has a macrostructure of 1 to 10s corresponding to injection, acceleration and extraction cycle. The beam extraction occurs within 20-40 ns bunches every 100-200 ns [40, 41]. Due to its relevance for this book, the Heidelberg Ion Therapy Center (HIT) synchrotron is described in a further sub-section.

2.3.2 Beam Delivery

Beam delivery can be divided into active beam delivery (e.g., pencil beam scanning, PBS) and passive beam delivery (e.g., passively scattered proton therapy).

Passive beams systems are fed with a narrow particle beam of $\approx 1\,\mathrm{cm}$. Then scatter foils are used to broaden the narrow beam to a given field. A rotating range modulator wheel is used to achieve a good tumor coverage (or distal conformation), the so-called SOBP. The final b eam s hape i s a chieved w ith m illed a pertures a nd c ompensators. The former shapes the treatment field to a desire target profile and the later shapes the distal part of the dose distribution [5, 40, 41].

Active beam delivery, such as PBS is more advanced, economical and precise than passive beam systems. In this technique, the beam is deflected by dipole magnets in overall

XY-axis (plane perpendicular to the beam direction). The *Z-axis* (axis colinear with beam direction) scan, and therefore the 3D conformity, is achieved with energy selection. The beam scan can be done spot-by-spot, by rastering, or in a continuous line. On balance, PBS allows a better target dose conformity, less beam contamination (e.g., lower number of neutrons) and requires less hardware [5, 40, 41].

PBS treatment plans use several energy layers and a finite number of spots per layer to conform the dose to the PTV. The treatments plans are computed with pencil-beam based algorithms and in some centers with *Monte Carlo* (more accurate but also more time demanding). In general, the spot separation is within 2 to 10 mm and 1 ms (magnets update) when considering time beam delivery [5, 40, 41]. The beam's position and energy are constantly monitored by multi-wire proportional chambers and ionization chambers, respectively.

2.3.2.1 Heidelberg Ion Therapy Center

The HIT, a unique facility in Europe, is the first in Europe to use both protons and heavy ions in treatments. It has started its operation in 2009 (with an active beam since 1997, GSI) treating more than 80000 patients with particle therapy since then.

The center is equipped with four beam stations (see figure 2.10), two with a fixed horizontal beam (H1 and H2 rooms), one with a gantry and the other one used for quality assurance (Q-A room), development and research activities. To produce the beam, the HIT uses a state-of-art accelerator chain consisted of two ion sources and an injector linear particle accelerator (LINAC) followed by a synchrotron [42, 43].

This book was carried out in the HIT. The institute provides an enormous va-riety of beam parameters for protons and ions, such as ^4He, ^{12}C and ^{16}O. Regarding pro-ton beams, HIT offers intensities from 8.0×10^7 to 3.2×10^9 particles/s and energies from 48.12 to 221.06 MeV/u. The spill macrostructure of the beam comprehends approximately a period of irradiation of 9 s with a duty cycle of 55 % [44].

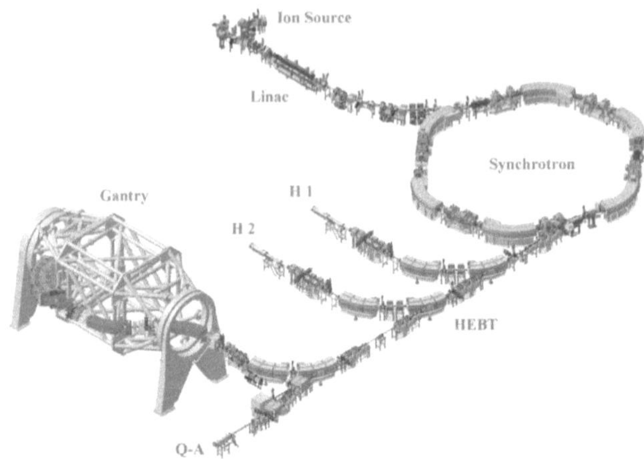

FIGURE 2.10: An overview of HIT accelerator: the ion source feeds the LINAC, where they get a first stage acceleration; followed by an acceleration stage synchrotron; next the beam is distributed by the high energy beam transport lines (HEBT) to the four beam stations, Gantry, H1 & H2 horizontal rooms, and quality assurance (Q-A) room [43].

2.3.3 In Vivo Range Monitoring

In vivo range monitoring is a hot topic, where *Bragg* peak uncertainty problems have been addressed in recent years. To be successfully applied in the clinical routine, it needs to satisfy certain points, such as [10]:

— can not interfere with the treatment beam nor with the patient;

— should be granty mountable;

— can not affect the treatment time;

— needs to be adequate to therapy facility features (e.g., beam time structures).

There are two significant lines of research in range monitoring technology, anatomic-tracking methods and beam-tracking methods.

2.3.3.1 Anatomic-tracking Methods

The first method to be successfully applied as in-vivo range monitoring was the positron emission tomography (PET). Particle-therapy PET measures the β^+ activity induced by the beam with a PET scanner. After a few seconds or minutes of beam patient irradiation, a scan is possible, allowing an online or offline measurement. Despite being a promising solution, the technique is limited by factors like biological washout (by metabolism, blood and lymph circulation), and blindness from PG [5, 7, 10, 45–47].

Proton radiography has been over research since the 1960s.It is performed by applying a high energy proton beam to the patient, which are detected on the patient's entry and exit. The residual range can then be directly measured, from which stopping power values of tissue can be directly calculated. Thus, proton radiography is a technique that can be used for treatment planning and/or range verification during off time beam. Despite its potential, proton radiography is not being used clinically and has a poor spatial resolution (if compared with conventional radiography) [5, 9, 10].

Magnetic resonance image (MRI) is another modality that can be used for anatomic-tracking of the beam dose deposition. Visible changes in tissue can be seen after patient irradiation. However, it was shown that by itself it can not verify proton range, and can only be used in a retrospective perspective [5, 8].

A more trivial way of anatomic-tracking is the use of implanted markers. Implanted markers rely on the possibility of implanting a dosimeter directly in the patient, ideally close to or within the tumor. Its use can provide $1D$ range signal with sub-millimetre precision (in vitro conditions) during or after each treatment field delivery [5, 48–50]. Nevertheless, a real implementation of the technique is limited by the fact that the implant can disturb the proton beam range. This is a major drawback that is seen as unviable clinically.

2.3.3.2 Beam-tracking Methods

Unlike anatomic-tracking methods, beam-tracking methods use indirect measurements to determinate the particle range. Beam-tracking methods are mainly based on PG emission, and in Chapter 3 it is presented a state-of-art regarding the subject. Notwithstanding, different techniques like beam track imaging by means of secondary-electron *Bremsstrahlung* and acoustic methods have been proposed [51–54]. Such techniques, however, lack of deeper research and are by far not mature for clinical trials.

Chapter 3

Prompt Gamma-Ray Imaging: State of the Art

The main advantage of PG is the ability to perform real-time verification of particle delivery. The PG can be detected instantaneously, within a few nanoseconds, following the nuclear interactions. Different groups have demonstrated the feasibility of the approach for monoenergetic proton pencil beams over the whole domain of clinically relevant proton energies.

The systems based on PG can be divided into imaging systems with physical or electronic collimation or non-imaging system. Features like energy, time-of-flight and position are exploited to retrieve the particle range or position of interaction. In Table 3.1 one can see a summary of the various features used by the different PG modalities. Furthermore, in the next sections, it is presented the description of different PG systems.

TABLE 3.1: Prompt-gamma modalities classified according to the prompt-gamma they exploit. Including in imaging systems are the pinhole, slit, multi-slit, knife-edge and *Compton* cameras. Some PG features are not mandatory. Adapted from [55].

PG features	imaging systems		non-imaging system		
	physical collimation	*electronic collimation*	*PG timing*	*PG peak integral*	*PG spectroscopy*
position	yes	yes	-	-	-
energy	optional	optional	optional	optional	yes
time-of-flight	optional	optional	yes	yes	optional

3.1 Single and Multi-Slit Cameras

The single-slit camera was first proposed and developed by Min et al. 2006 [56–58].
It is composed of a collimator with a slit. The collimator has the objective of moderate and
capture fast neutrons before reaching the detector system (scintillator + photon detector).
The camera allows axial measurements of emitted PG by spot. Min et al. observed that the
number of PG emissions decreased for measurements beyond *Bragg* peak [56]. Therefore,
demonstrating the correlation between the emission of PG and particle range. This system
was also verified has proof of concept for carbon particles [59].

Single-slit cameras only allow single position measurements, and then the system
needs to be moved to make another measure. The need for a mechanical moving pre-
vents the use of the technique clinically.

In such a scan procedure is not clinical feasibility. Thus, the slit camera was extrapo-
lated to a multi-slit camera, where several slits and system detectors are used [55, 60, 61].
An illustration of the system can be seen in figure 3.1 [33]. In figure 3.1 it is possible to
see some extra system. The hodoscope is useful to get a 3D distribution of the PG emissions and allows time-of-flight measurements. Min et al. 2012 reported a good correlation
with the range using the multi-slit camera [62]. This system was also verified for carbon
particles [33]. However, its accuracy has not been verified experimentally for clinical ap-
plications [55]. Notwithstanding the lack of experimental proof, more recently, Pinto et
al. 2014 trough *Monte Carlo* simulations reported how multi-slit cameras could achieve a
clinical scenario [61].

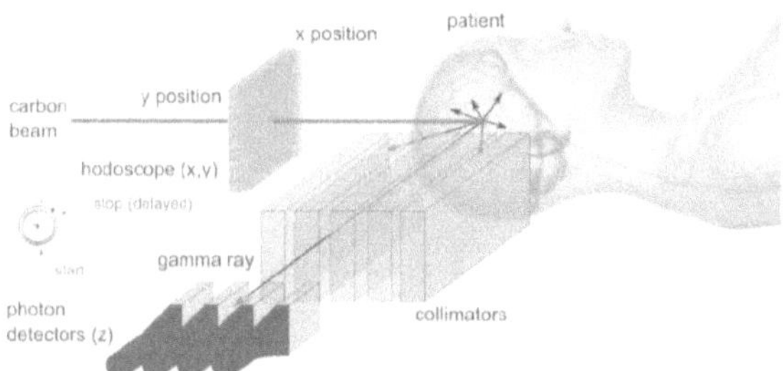

FIGURE 3.1: Illustration of a multi-slit camera system. The hodoscope allows measur-
ing the particle position in the sagittal and coronal view. Additionally, it gives information about time coordinates. The system allows axial measurements of PG emissions.
Adapted from [33].

3.2 Knife-Edge Slit Camera

An alternative to the multi-slit camera is the knife-edge slit camera. First introduced a pinhole camera by Kim et al. 2009 [63], the knife-edge slit camera is based on the physical principles of classical optics adapted to PGI. The system concept is illustrated in figure 3.2. The knife-edge slit allows measurements of PG emission along the longitudinal axis. The first studies using knife-edge slit camera reported a very good sensitivity to the particle ranges with standard deviation values of $\approx 2\,\text{mm}$ and range shifts detection lower than $2\,\text{mm}$ [64, 65]. The first PGI clinical application was achieved with a knife-edge slit camera in passive beam mode [66]. The research results shown range variations of only $\pm 2\,\text{mm}$. Later, Xie et al. 2017 reported the first clinical results with PBS [67]. Simulated results show that the accuracy of the knife-edge camera can reach values of $\approx 1\,\text{mm}$ [68]. More recently, Berthold et al. 2020, demonstrated, with in-vivo measurements, that the system was able to detect strong anatomical changes in prostate cancer treatments [69].

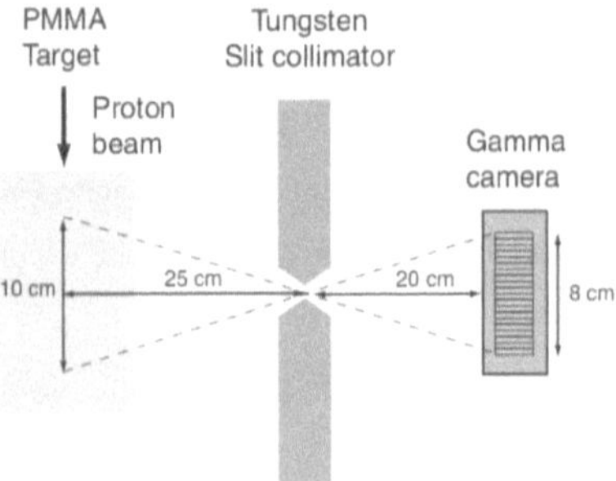

FIGURE 3.2: Schematic of a knife-edge slit ('Tungsten Slit collimator') system. Adapted from [70].

3.3 Compton Camera

The previous PGI techniques use heavy collimators that can be an obstacle to clinical feasibility. An alternative to this systems is the *Compton* camera. Its main advantage is the use of electronic collimation instead of a physical one. Additionally, due to this type of collimation, they have a higher detection efficiency [19, 33]. Commonly, *Compton* cameras consist of multiple position-sensitive PG detectors [38]. At least one scatterer detector and one absorbed is needed. However, the use of several scatterer planes is an option. The aforesaid system can be represented by image 3.3. The energy- and position-resolved detection from the multiple detectors restricts the direction of incident PG in a cone. The origin of the photon can then be determined with the vertex superposition of multiple cones [5, 10, 38]. The *Compton* cameras have the potential to get image proton dose in 3D. Drawbacks that limits its use in a clinical scenario are its complexity, expensive electronics, low coincident efficiency, high detector load and a high percentage of random coincidences [38].

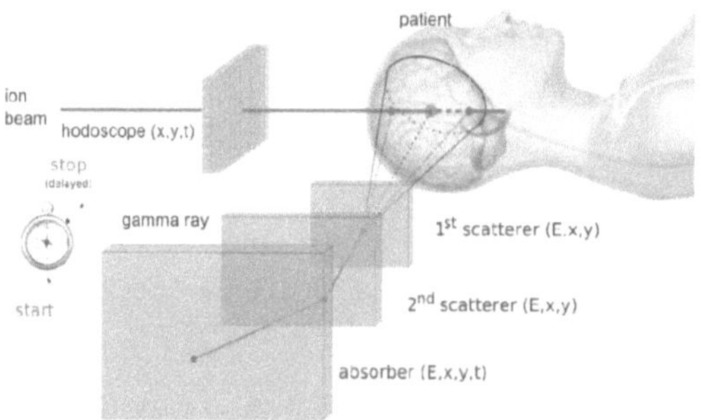

FIGURE 3.3: Schematic of a *Compton* camera system. The hodoscope allows to measure the particle position in the sagital and coronal view. Additionally, gives information about time coordinates. Adapted from [33].

3.4 Prompt Gamma-Ray Timing

The prompt gamma-ray timing (PGT) technique uses the proton transit time to infer about its range. The transit time is measured indirectly by the temporal prompt gamma-ray emission, thus determining the particle time-of-flight (TOF). For such, usually, it is used a time reference provided by the accelerator particle bunch radio frequency (RF). Then range is derived mathematically through the CSDA theory. To be more efficient, PGT only uses gamma energies between 3 MeV and 7 MeV, allowing neutrons background rejection [38, 55, 71].

In figure 3.4, it is presented a TOF spectrum example [55, 72]. Mathematically, the statistics parameters, peak mean, and variance are the ones that provide transit time information [38, 55, 71]. As observed by figure 3.4, the peak width increases with the energy, and as expected, the peak mean shifts towards a lower time-of-flight for high energies.

PGT concept was first introduced by Golnik et al. 2014, where he demonstrated that PGT statistics is dependent on protons transit time in a target [22]. Later, Hueso et al. 2015 showed detection of 5 mm and 2 mm range shifts for low and high statistic, respectively [72].

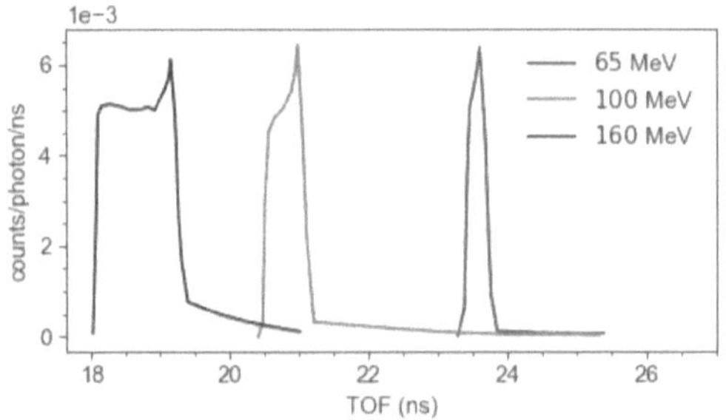

FIGURE 3.4: TOF spectrum, the peak width increases with proton velocity/energy. Adapted from [55].

A PGT problem is the accelerator bunch time drift against the time reference RF. With that in mind, Petzoldt et al. 2016 developed a proton bunch monitor to make an independent time reference [73]. With two detectors working in coincidence, the bunch monitor can detect scattered particles in a hydrogen-containing foil (placed at nozzle exit). Thus, making it possible to do background correction of uncorrelated events. Paush et al. 2016 suggested the improvement of system detection to achieve better time resolution and stable PGT measurements [71]. Additionally, and adding more detectors as proposed by Werner et al. 2019, PGT statistics can be improved [74]. Recently, and using ultrafast fast detectors, Marcatili et al. 2020 described the possibility to monitor a pencil beam range spot-wise [75].

3.5   Prompt Gamma-Ray Peak Integral

Prompt gamma-ray peak integration (PGPI) is a technique proposed by Krimmer et al. 2017. Intended to be used as a simple and independent monitoring device, PGPI exploits the integral in prompt gamma-ray TOF distributions. Thus, determination of the *Bragg* peak position can be obtained from prompt-gamma count rate ratios. Although never used in a clinical scenario, it has the potential to achieve a range precision of 3 mm with only 10^8 incident protons. Additionally, and using combination signals of multi detectors, it is possible to detect target misplacements [10, 40].

3.6 Prompt Gamma-Ray Spectroscopy

Prompt gamma-ray spectroscopy (PGS) uses the intensity ratios of prompt gamma-ray characteristics lines to infer the beam energy and the residual range of the particles (via energy dependence of the cross-sections). Furthermore, with PGS, it is possible to get information about the target composition. Thus, one can measure changes in levels of tumor hypoxia, and therefore, tailor the treatment to needs [10, 55].

In figure 3.5 it is possible to see PGS results of intensity ratios versus point depth of measurement. The individual intensities ratios result in a monotonic and unequivocally invariable data, therefore demonstrating that the particle absolute range can be retrieved analysing the prompt gamma-ray spectrum [13].

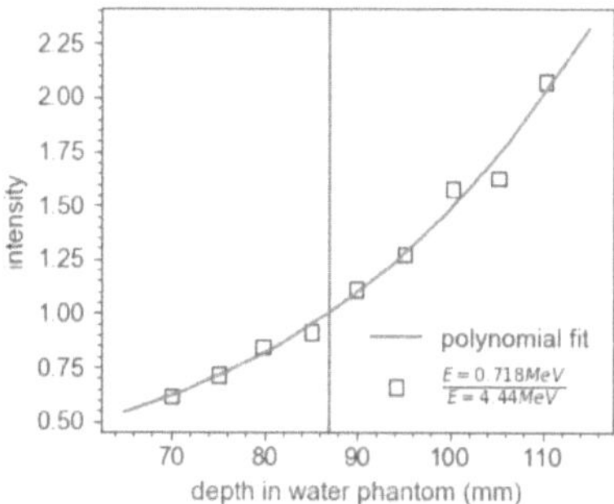

FIGURE 3.5: The ratio between the intensities of one pair of PG lines. The measurements were taken at different distances in depth. Green curves are the polynomial fit of the experimental data, and the vertical red lines indicate the position of the *Bragg* peak. Adapted from [13].

Verburg and Seco carried out the first direct studies regarding PGS. They showed that PGS has the capacity to infer about the absolute range of the protons [12, 76–78]. Later, with a more robust technique, Dal Bello and Seco, and Martins and Seco continued the previous works [13, 14, 44, 79–81]. New mathematical models were developed with Monte Carlo simulations, and a combination of energy and time-resolved led to a substantial improvement in determining the particle range using PGS. Meanwhile, Martins et al. 2020 developed a system with a beam trigger based on scintillating fibers demonstrating the possibility of resolving time within sub-nanosecond scale [44]. Moreover, Dal Bello et al. 2020 showed, for the first time, that it is possible to correlate the Bragg peak position with features extracted via PGS for carbon beams [13].

Issues like statistics per prompt gamma-ray line and mixed beams with target heterogeneities are a subject of study that still needs to be addressed in PGS [55]. Nevertheless, and regarding clinical scenarios, a prototype was developed by Hueso et al. 2018, where he demonstrated a statistical precision of 1.1 mm for absolute proton range [11]. More recently, Martins et al. 2020 showed the importance and how PGS can be used to measure oxygen and calcium concentrations to the assessment of tumor hypoxia and tracking of the calcifications in brain metastases [14].

Chapter 4

Proton Therapy in Prostate Cancer

Prostate cancer is one of the most common cancer in men [82]. The most common treatment for prostate cancer is surgery and photon therapy. Nevertheless, proton beam therapy (PBT) has been increasing and already targets a significant number of patients ranging from 2 % to 6 % when compared to photon therapy. Several aspects make PBT a desirable choice for prostate cancer, among them [39, 82]:

— high conformity (lower penumbra for intermediate depths) and effectiveness;

— reduce the dose to normal tissues due to *Bragg* peak distal falloff;

— less radiation-induced secondary cancers compared to photon therapy;

— PBT is ideal for dose escalation (improving biochemical progression-free survival).

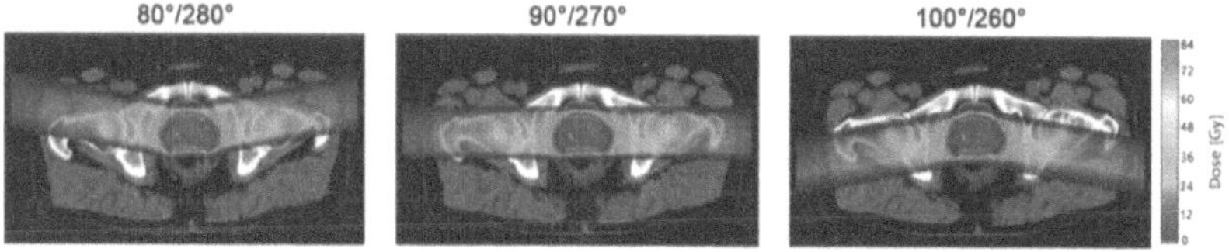

FIGURE 4.1: Dose distribution for three beam angle configuration, 80°/280° (anterior oblique beams, left image), 90°/270° (lateral beams, central image) and 100°/260° (posterior oblique beams, right image). Adapted from [83].

Due to its particular location, prostate cancer, is surrounding by multiple organs at risk (OARs) that need to be taken into account. Particular attention is given to the rectum and bladder, where late toxicities have been observed and are a limiting factor in prostate cancer treatment (e.g., the rectum wall can be exposed to more of 15 % of the 95 % of total dose) [2, 4]. Also, PBT in prostate cancer has a higher sensitivity to target movement (coverage lose), additional margins (transverse and longitudinal) and inter- and intra-fraction movements with target displacements of 4 mm to 5 m, and 0.1 mm to 24 mm, respectively [2, 39, 84].

Several investigations have been carried out regarding better techniques in PBT for prostate cancer [15, 48–50, 82–89]. In the next sections is made an overview of these researches.

4.1 Bilateral Beams

PBT for prostate cancer, usually, it is done with lateral beams (BL). The BL setup 4.1) is intentionally used to aim the lateral penumbra ($\sim$ 10*mm* width at 25*cm* range) to the anterior rectum wall [82, 89]. Thus, avoiding the distal end of spread-out *Bragg* peak ($\sim$ 4*mm* width at 25*cm* range) and the uncertainties like longitudinal motion, CT number accuracy, conversion to proton stopping power and penetration depth [39].

Additionally, BL tends to be more robust. Nevertheless, prostate cancer treatment with BL still has problems, such: high doses to healthy tissues, target or normal tissue constraints are not always fulfilled, relies on lateral penumbra that is much broader than the distal falloff and can not be adequately applied in patients with a hip prosthesis.

4.2 Anterior Oblique Beams (and Posterior Oblique Beams)

The spread-out Bragg peak and its capacity of spare adjacent organs at risk can be exploited with the use of anterior (AO) or posterior oblique (PO) beams (see figure 4.1). The main advantages of using AO/PO beams are lower mean doses in organs at risk (rectum, penile bulb and femoral heads), possible avoidance of hip prosthesis, better potential for dose escalation (hypofractionation, increasing of tumor control probability), and better ability to conform radiation dose to tumor [85, 89].

AO configuration increases the bladder dose and decreases the rectum dose; the opposite is seen for PO configuration [83]. As a result, AO configuration for the rectum has higher NTCP compared to the outermost PO configuration [88, 89]. Overall, with AO/PO

beams, there is a potential to avoid common toxicities such, urinary toxicity, rectal procti-
tis, gastrointestinal toxicity, erectile dysfunction or hip pain in organs at risk surrounding
the prostate [15, 82, 89].

On the other hand, some studies have been pointed out problems regarding the application of AO/PO beam. Concerns about inter- and intra-fraction movements are the
most critical. Anatomy variations in target region due to these movements increase the
hazard of under dosage of the prostate or/and overdosage of surrounding healthy tissues
[15, 39, 82, 84, 89, 90]. Another problem is related to linear energy transference (LET) [83].
As LET increases toward the distal end of spread-out *Bragg* peak, so areas of higher RBE
(hotspots) towards OARs can be expected [82, 83, 87]. In order to mitigate these problems
is suggested and often used rectal balloons and rectum-prostate spacers in PBT of prostate
cancer. Alternatively, it is recommended in-vivo proton range verification on a daily basis, allowing margins to be significantly reduced, and if needed, the patient re-evaluation
planning [87, 89, 90].

4.3 Rectum-Prostate Spacers

If the space between the rectum and prostate is increased, the dose to the rectum
decreases. To achieve such type of condition, a hydrogel spacer between the prostate
and rectum can be strategically placed (see figure 4.2) [15, 84, 87, 88]. Such configuration
allows a better margin control and avoids high LET regions in the anterior rectum wall
(dose escalation in the rectum can be mitigated).

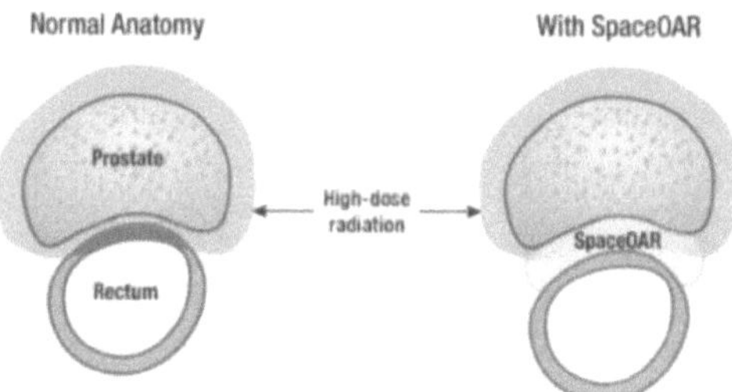

FIGURE 4.2: Representative diagram of a hydrogel spacer and how it creates space between the rectum and prostate. Adapted from [91].

With no spacer, the prostate-rectum mean distance reported by Chung 2017 et al. is 1.5 mm while with the spacer, the distance is increased to 9 mm [88]. Considering margins of $3.5\% \times beamrange + 2$ mm, his study showed a decrease of $V70_{rectum}$ from $5\% < V70 < 18\%$ to $0\% < V70 < 5\%$ with the use of a spacer [88]. Despite the advantages of using spacers, it is needed to keep in mind that asymmetric/non-homogeneous insertions could lead to hotspots in rectal biological dose, gel migration can happen, it is an invasive technique, and last but not least, low-grade symptoms (e.g., mild tenesmus/dyschezia or rectal ulcer) can occur [15, 87].

4.4 Rectal Balloon

A way of inter- and intra-fractions movement mitigation is the use of a rectal balloon. Usually filled with water when inserted in rectum can fix the prostate position and lower the anatomy variations between or during fractions [15]. This type of application in PBT of prostate cancer is only used with BL due to a lack of in-vivo range verification techniques. Recently, and as a complement of the rectal balloon, metal oxide semiconductor field-effect transistors (MOSFETs) and diodes have been used as a range verification technique [48–50, 86, 89]. The last, showed promise results when submitted to an in-vivo validation and commissioning [49, 86].

Chapter 5

Materials and Methods

In the chapter, it is described in all the materials and methods used experimentally. All targets and materials used are described, followed by a brief detail of the acquisition system's main components. Finally, some results acquired with the system and how the data processing was conducted are presented.

5.1 Targets & Materials

5.1.1 Flasks & Samples

Polystyrene flasks filled with different samples were used in preliminary tests (see figure 5.1). The flasks had a wall thickness of $l_{wall} = 0.15\,\text{cm}$ and an inner cavity of $l_{cavity} = 3.2\,\text{cm}$. Three types of samples were used: double distilled water (H_2O); double distilled water mixed with silicon dioxide ($H_2O + SiO_2$) in a ratio of 1 to 1 and 3 to 2 (e.g.: 90 ml of H_2O to 60 ml of SiO_2 for the 3 to 2 ratio); and commercial silicone sealant.

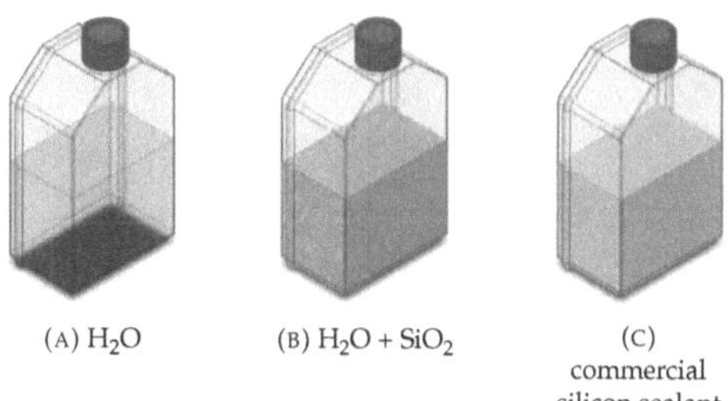

FIGURE 5.1: Flasks representation. The flasks were filled with different samples and used as targets in this book.

5.1.2 Prostate Phantom

In order to carry out in vitro experiments (close to clinical reality), a realistic and durable prostate training phantom *CIRS Model 070L* (see figure 5.2) was used. The phantom includes tissues such as the bladder, rectal wall, perineal membrane, prostate gland and seminal vesicles. All tissues were contained in a box of 9 cm × 10 cm × 10 cm filled with a Zerdine® gel tissue-equivalent. Zerdine® gel is a patented material based on a mixture of solid-elastic, water-based polyacrylamide [92]. It can be used for computed tomography (CT) scans, and the formulation can be adjusted to mimic a variety of soft-tissues [92].

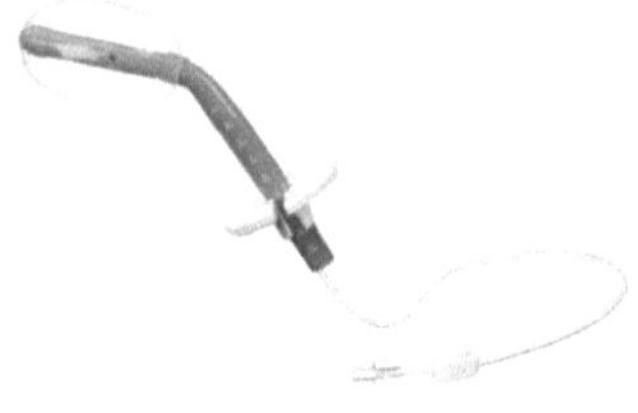

(A) Prostate training phantom *CIRS Model 070L*   (B) RectalPro™75 Endo Rectal Balloon

FIGURE 5.2: Representation of the prostate training phantom *CIRS Model 070L* (left) and RectalPro™75 Endo Rectal Balloon (right) [93, 94].

5.1.3 Rectal Balloon

Additional to the prostate phantom, it was used a *RectalPro™75 Endo Rectal Balloon* (see figure 5.2) filled either with 50 ml of H_2O or 50 ml of H_2O + SiO_2. The rectal balloon is designed as an immobilizer to assist in positioning the prostate in a more predictable and reproducible location during computed tomography (CT) and treatments. In figure 5.3 it is possible to see a CT of the system phantom + rectal balloon. The CT was taken in HIT with a single energy scan.

ImageJ software was used to measure the balloon diameter, $d_{balloon} \approx 4$ cm and the gap between the balloon and the prostate, $d_{gap} \approx 0.5$ cm (see 5.4). The measures were taken manually at 50 % of the slope transition between structures. The gap values between the rectum wall and the prostate are higher than the mean values found in literature [88]. However, this discrepancy does not influence future conclusions.

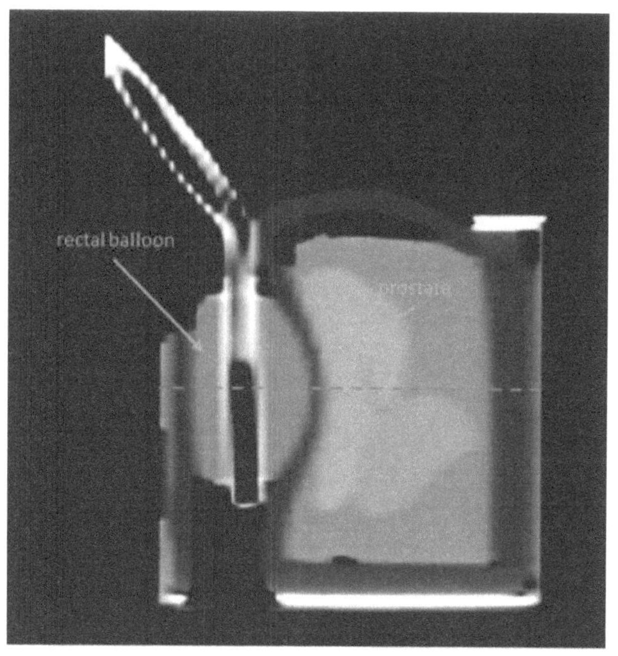

FIGURE 5.3: Image of a CT of the prostate training phantom *CIRS Model 070L* (sagittal view). The blue dashed line indicates the trace profile taken for the figure 5.4.

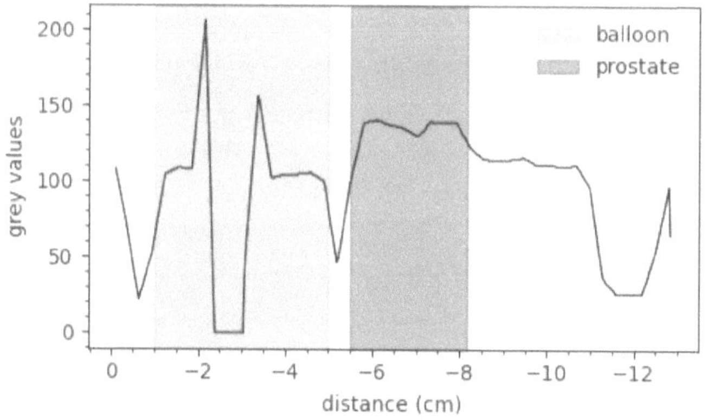

FIGURE 5.4: CT grey value profile taken over the blue dashed line of figure 5.3.

5.2 Acquisition System

5.2.1 Primary Detector

The primary detector is a $CeBr_3$ hygroscopic scintillating crystal optically coupled to a *Hamamatsu* photomultiplier tube (PMT) *R9240-100* (see figure 5.5). It has a decay time of 20 ns and a density of $5.1\,g/cm^3$. The crystal, protected by Teflon and aluminium, has a cylindrical shape with a diameter $d = 3.81\,cm$ and length $l = 7.62\,cm$ When compared with other crystals used in PGI, such as $LaBr_3$, the $CeBr_3$ as similar characteristics but a lower background. All raw data was obtained with primary detector.

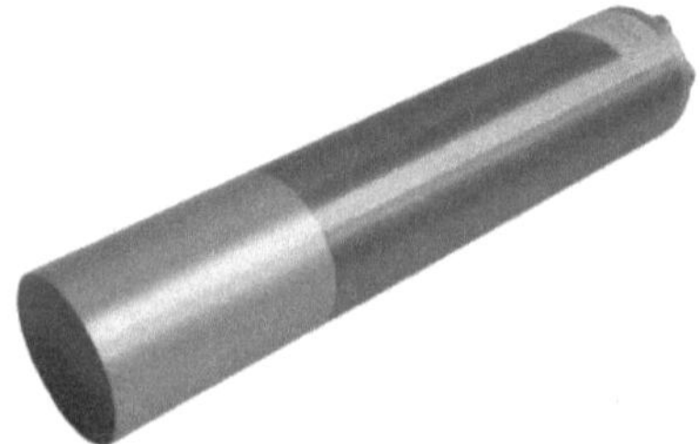

FIGURE 5.5: Primary detector.

5.2.2 Secondary Detector

The secondary detector is constituted by bismuth germanate crystals (*BGOs*) sectioned in eight optically separated and azimuthally symmetric segments. Each section is optically coupled to an independent *Hamamatsu* PMT *R1924* followed by a pre-amplified circuit. BGOs crystals have a decay time and a density of 300 ns and $7.13\,g/cm^3$, respectively. Due to its high effective atomic number and density, it has a better gamma absorption when compared with $CeBr_3$. The eight individual components are contained in a cylindrical shape and hollow cylinder to fit the $CeBr_3$ (see figure 5.6). The *BGO* with energy- and time-resolved signal was used as an anti-coincident shield (AC), allowing the background suppression of *Compton* and single/double escape events. The time resolution between the $CeBr_3$ and AC is $FWHM = 3.58\,ns$ [80].

<table>
<tr><td align="center">(A) photo</td><td align="center">(B) detailed</td></tr>
</table>

FIGURE 5.6: Secondary detector and a draw showing the eight individual *BGOs* crystals.

5.2.3 Beam Trigger

To enable time-of-flight measurements, the prototype is equipped with an array of scintillating fibers (beam trigger) with a decay time of 3.2 ns and a subnanosecond time resolution of 0.8 ns [44]. The fibers (*BCF-12* from *Saint Gobain Crystals*) have a diameter of 0.5 mm and are agglomerated in a cross matrix of sixty elements connected to two *Hamamatsu* PMT *R657* ($4 fibers/PMT$). The fibers and PMTs are enclosed in a light-shielding box with an external window for the beam (see figure 5.7). The window is made of aluminized mylar with a thickness $\approx 10\,\mu m$. The beam trigger provides time information to derive the TOF spectrum allowing the background suppression of uncorrelated events (e.g., hydrogen neutron capture).

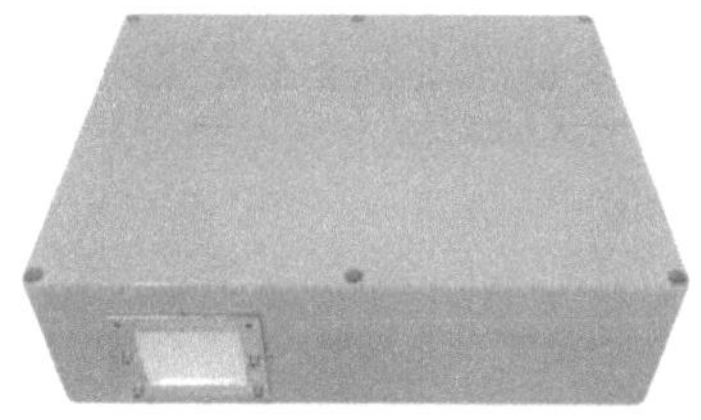
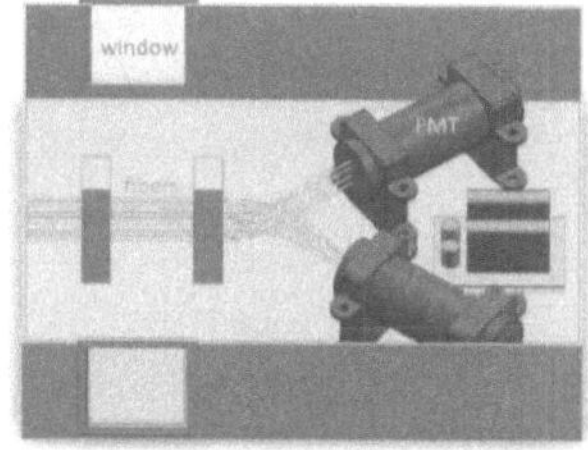

<table>
<tr><td align="center">(A) photo</td><td align="center">(B) detailed</td></tr>
</table>

FIGURE 5.7: Beam trigger enclosure and a 3D detail view of its inside.

5.2.4 *FlashCam FADC* **module**

The PMTs readout was performed with a high-performance ADC, *FlashCam FADC* (see figure 5.8). The *FlashCam FADC* system has a multi-channel acquisition mode capable of digitalizing data at 250 MS/s with 12-bit resolution. It has independent channels that perform a readout of all channels as soon as a trigger is activated. Its connection to a computer is done with a flexible ethernet-based interface.

All acquisitions in this book were performed event-by-event with single traces of $\Delta t = 240$ ns and sampling intervals of $\delta t = 4$ ns for primary/secondary detector and $\delta t = 1$ ns for the beam trigger.

FIGURE 5.8: *FlashCam FADC* module used to digitalize all PMT signals.

5.3 Data Analysis Procedure

The analysis algorithms were all written in C++ and *Python* languages. Regarding python language, in this book, the main libraries used were *numpy, pandas, matplotlib, scipy, sklearn* and *pickle* [95–100]. The next sub-sections are based on data processed in the Chapter 6.

5.3.1 Raw Data Processing

5.3.1.1 Signal Fit (Photomultiplier Tubes)

As mentioned before, the PMT data is read by the *FlashCam FADC* module. The output trace for each event and PMT is then processed. This process implies a baseline restoration, and a fit for each peak found (maximum of 3 peaks per t race). Three types of fits were tested, a *Gaussian* fit, a log-normal fit, and an exponential modified *Gaussian*

(EMG) fit. The best results were obtained with the last one. In figure 5.9 is presented a trace event with an exponential modified normal fit (EMG).

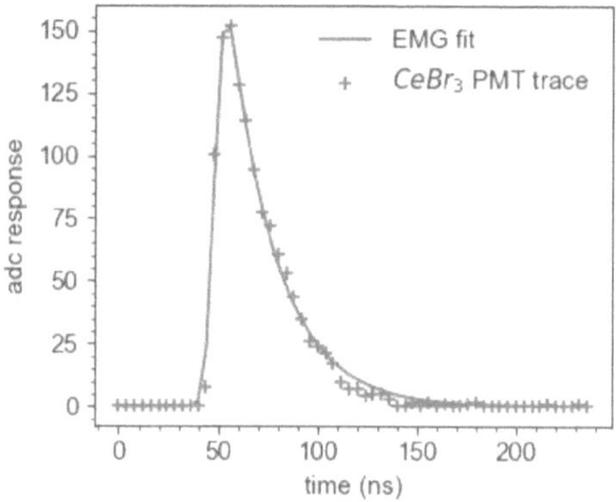

FIGURE 5.9: An event trace (blue points) acquire with *FlashCam FADC* module and fitted with an EMG.

The EMG distribution combines a *Gaussian* with an exponential distribution and its mathematical expression can be given by the equation 5.1, where t is the time, a is the *Gaussian* amplitude, σ is the *Gaussian* sigma, μ is the position of the unmodified *Gaussian*, γ is the relaxation time parameter of exponent used to modify *Gaussian* and $erf = \frac{2}{\sqrt{\pi}} \int_0^z e^{-t^2} dt$ [101].

$$f(t,a,\mu,\sigma,\gamma) = \frac{a\gamma}{2} e^{\left[-(t-\mu)^2/2\sigma^2\right]} \left\{ 1 + erf\left[\frac{\mu + \gamma\sigma^2 - t}{\sqrt{2\pi}}\right] \right\} \tag{5.1}$$

For each peak, the detected informations on the event identification, a, μ, σ, γ, area (A), mode (m), height (max), and adjusted R-squared ($r^2_{adjusted}$) were stored. The area, mode and height values were obtained with numerical calculation while the adjusted R-squared was derived from the coefficient of determination. For figure 5.9, the calculated parameters values are in table 5.1.

TABLE 5.1: Calculated values for the fit of figure 5.9.

fit parameters			
a	5.07×10^3	A	5.06×10^3
μ	4.82×10^1	m	5.39×10^1
σ	3.56	max	1.56×10^2
γ	4.15×10^{-2}	$r^2_{adjusted}$	9.85×10^{-1}

5.3.1.2 Valid Events Filtering

Not all detected peaks are considered as valid events. Some of the events had a bad shape (e.g., due to overflows or pile-ups) and others were simply uncorrelated.

For each run, the distribution of the adjusted R-squared was computed. Then, a threshold ($thr_{r^2_{adjusted}} = \mu - 7.5\sigma$) was defined. The threshold could be a simple constant value, but the former approach was used for a more consistent selection. In figure 5.10, it is possible to see the result of such mathematical process, where all events above the threshold are valid vents, and the ones below are removed from the dataset.

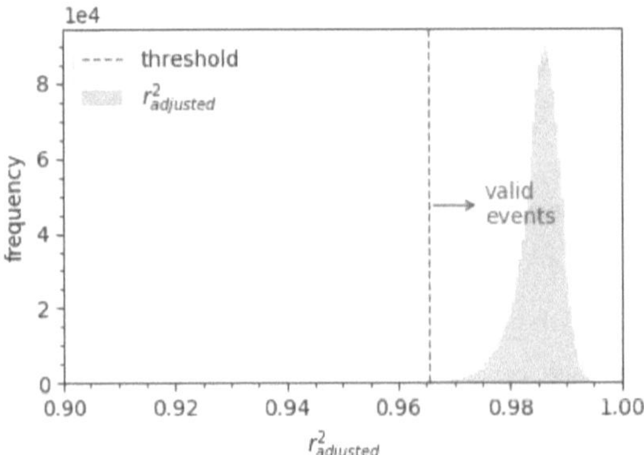

FIGURE 5.10: Distribution example of $r^2_{adjusted}$. Only the events above th threshold were selected for the PG spectrum.

The spill is defined in this book as the time which the beam is active. The time of spill-on is correlated with the trigger event time. Thus, analysing the triggers events time, it is possible to determine the number of spills and the time of each spill. The number of spills used for singles spots analysis was 14. In figure 5.11 and table 5.2, it is shown, as an example, the result of the spill analysis. The spill identification was achieved through derivation and threshold cut. From the table 5.2 one can observe that the spill time $\approx 5\,\mathrm{s}$. The value is within the values reported by Martins et al. 2020 [44].

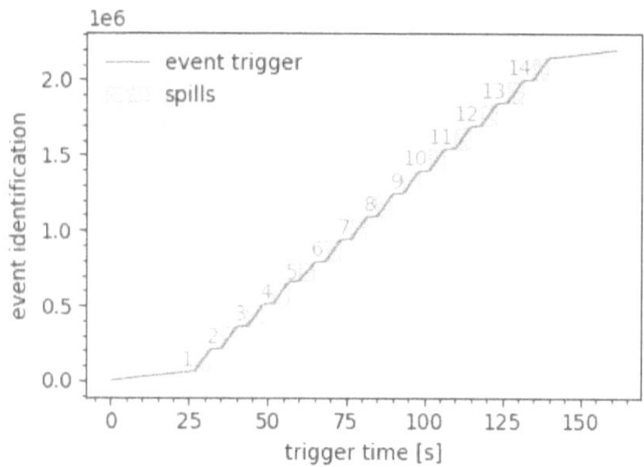

FIGURE 5.11: Events as function of their trigger time (absolute value). The green areas are the events within the spills.

TABLE 5.2: Initial, final and absolute time of each spill.

spill number	initial time (s)	final time (s)	span (s)	spill number	initial time (s)	final time (s)	span (s)
1	26.81	31.78	4.90	8	84.76	89.75	4.99
2	35.11	40.09	4.98	9	93.06	48.04	4.98
3	43.39	48.38	4.99	10	101.5	106.3	4.80
4	51.59	56.57	4.98	11	109.7	114.6	4.97
5	59.88	64.87	4.96	12	117.9	122.9	4.98
6	68.63	73.16	4.52	13	126.2	131.2	4.98
7	76.47	81.45	4.97	14	134.5	139.5	4.97

5.3.1.3 Dead Time Correction

The dead time in count-based detectors is defined as the time after each event in which the detector can not record another event. Detectors can be characterised regarding their behavior as non-paralyzable or paralyzable systems. In non-paralyzable systems, the detector is blind during the acquisition time and therefore, all new events that reach the detector (within acquisition time) are lost. On the other hand, in paralyzable systems, when a new event occurs (within acquisition time) a new measurement is triggered and, consequently, the dead time is restarted [18].

The *FlashCam FADC* module is a non-paralyzable system and its dead time behaviour is given by equation 5.2, where N is the real count rate, N_m the measured count rate, T the time interval and τ the dead time [18]. A dead time measurement follows each trace acquisition. Thus, with the dead time information and equation 5.2 it was possible to calculate the dead time corrections. Note that during all experiments the dead time was always $< 15\%$ for intensities in ranges of 10^8 to 10^9 protons/s (and most of the times *deadtime* ≈ 0).

$$N \approx \frac{N_m}{1 - \tau/T} \tag{5.2}$$

5.3.1.4 Spectra Filtering

A PG spectrum can be obtained by means of a histogram. For such, the values of parameter *a* obtained in the fit process were used. In figure 5.12, one can see a PG spectrum before and after filtering. The parameter *a* has a threshold between 5×10^3 (lower bound) and 2×10^4 (upper bound), which cuts most of the undesired events. Uncorrelated events (e.g., β^+ decay process) in the low energy spectrum are also filtered.

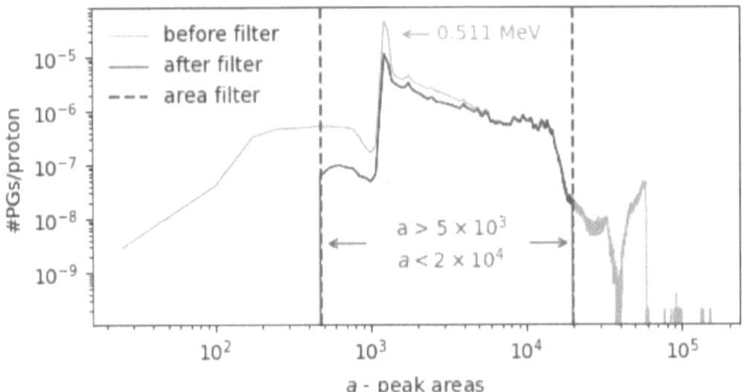

FIGURE 5.12: PG Spectrum before and after filtering. Vertical green lines indicate a delimited threshold filter in peak areas (*a* parameter). β^+ decay and consequent annihilation photon production (corresponding to 0.511 MeV energy) is pointed out by a grey arrow.

5.3.1.5 Spectra Denoising

Errors in the PG spectrum analysis may arise due to noise phenomenons such as detector statistics fluctuations or electronic system noise. Therefore, the noise signal needs to be mitigated before any qualitative and quantitative analysis. A standard algorithm used for the smooth is the *Savitzky–Golay filter*. Based on the least square method, the *Savitzky–Golay filter* can preserve the spectrum peaks information while eliminates the noise. A zoom-in of the previously filtered spectrum is depicted in figure 5.13. The smooth parameters, namely the polynomial order and window length, were optimized for each analysis in this book.

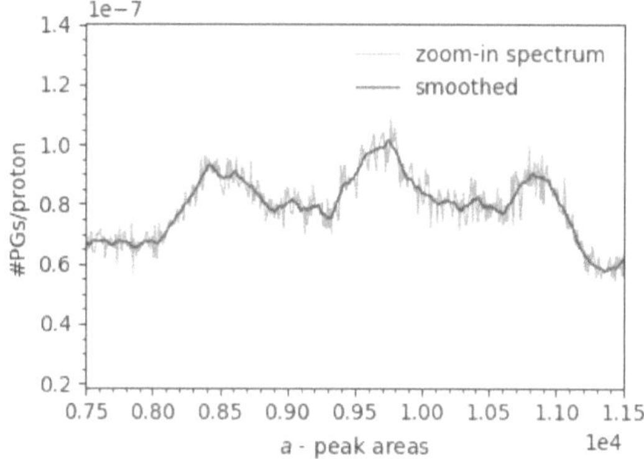

FIGURE 5.13: Zoom-in of the spectrum of figure 5.12.

5.3.1.6 Spectra Calibration

Detector calibration allows PG spectra to be interpreted in terms of energy rather than other meanless quantity (e.g., analog-to-digital channel). There are several sources of calibration errors. An example of such errors is the energy shifts due to source/detector orientation and distance. PG spectra measured in human tissue or water equivalent tissues allow for a self-calibration of the detector, where PG lines of oxygen and the annihilation (from β^+ decay) peak can be used for spectrum calibration. An example of a linear regression calibration and a calibrated PG spectrum are depicted in figure 5.14, and 5.15, respectively.

Figure 5.15 is labeled with some important features. The identified peak, 0.511 MeV, happens due to the detection of photons from β^+ decay and is the most prominent peak in the spectrum. As an example, it is also pointed out a PG line of the ^{12}C de-excitation (see appendix A) and the corresponding SE and DE peaks. Finally, and next to the 2.31 MeV peak, one can find a peak with the energy of 2.22 MeV associated with the hydrogen neutron capture. In chapter 6, this and others PG spectra will be the subject of further study.

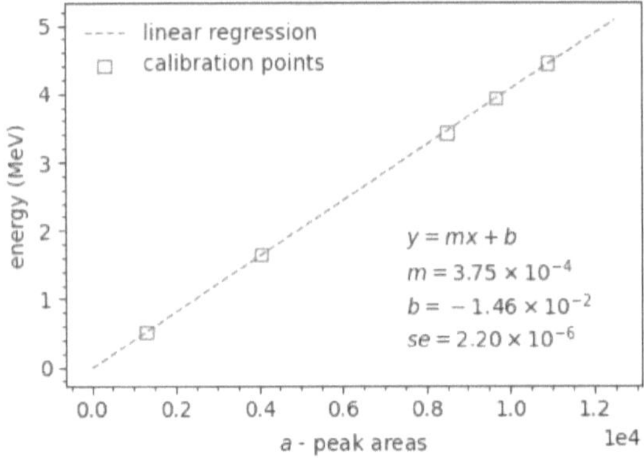

FIGURE 5.14: Calibration curve for the detector system used in this book. Calibration based on known oxygen lines of a standard PG spectrum. In the linear regression parameters box, *se* stands for standard error.

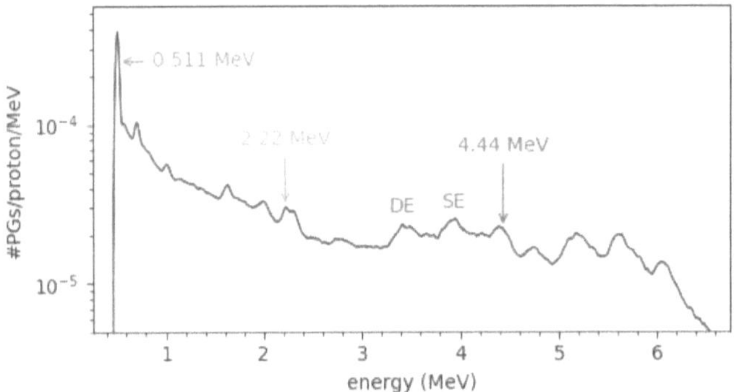

FIGURE 5.15: A calibrated PG spectrum example. Annihilation peak (from β^+ decay, 0.511 MeV), hydrogen neutron capture peak (2.22 MeV), and ^{12}C de-excitation peak (4.44 MeV) are pointed out by grey, yellow and red arrows, respectively. Next to ^{12}C de-excitation peak are also identified the correspondent single and double escape peaks.

5.3.2 Background Suppression

The background suppression was based on primary, secondary, and beam trigger detectors data. Two techniques were used for background suppression: time-of-flight and anti-coincidence shield.

5.3.2.1 Time-of-Flight

A typical time-of-flight spectrum (based on book data) is represented on fig-ure 5.16 (a). Three regions can characterize a TOF spectrum in PGS: prompt gamma-ray (S_{PG}) region, scattered radiation region and time-independent radiation region. S_{PG} in figure 5.16 (a) is shown as a green area and is defined by a cut at 2.5σ of a basis *Gaussian* fit (red dashed line) to the PG peak. Such cut, results in a window of $\Delta t_{ToF} \approx 10$ ns. In this book, the Δt_{ToF} was set with calculations on-the-fly for each PG spectrum. The obtained values were in agreement with those found in the literature [13, 74]. Scattered radiation area (due to *Compton* events) is next to S_{PG} area (before and after) and is usually defined with a small window of $\Delta t_{scattered} \approx 5$ ns [13]. Time-independent radiation, like hydrogen neutron capture and β^+ decay radiation, largely happens outside of the previ-ous windows. Better insight into how TOF relates with PG spectra is represented on the figure 5.16 (b). The PG region of emission and some energy-resolved lines can be easily identified. Additionally, for lower energies, it can be noted that PG information is more spread out over the TOF. This effect is associated with the high probability of scattered effects for low energies.

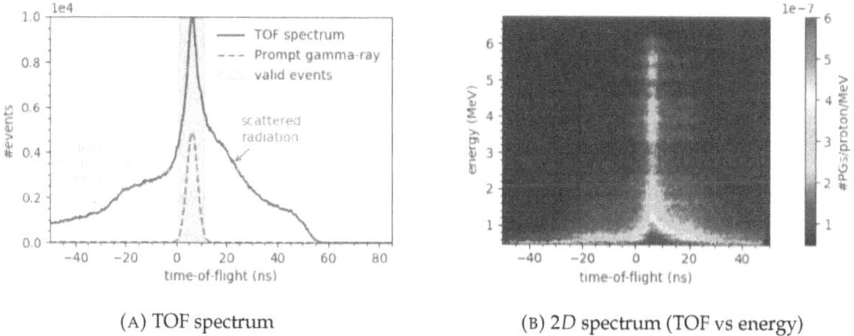

(A) TOF spectrum

(B) 2D spectrum (TOF vs energy)

FIGURE 5.16: TOF spectrum. Prompt gamma-ray (red dashed line) events were fitted with a *Gaussian* curve. A cut for valid events were defined at 2.5σ ($\approx 99\,\%$) corresponding to a window $\Delta t_{ToF} \approx 10$ ns.

One can see, in figure 5.17, the effect of a background suppression based on TOF measurements of figure 5.14. As expected, the background is significantly reduced due to the TOF background suppression technique. The yellow arrow in figure 5.17 serves as a marking for the hydrogen neutron capture peak. As mention before, hydrogen neutron capture is a PG time-uncorrelated process. Thus, applying the TOF technique allows its mitigation in the PG spectrum.

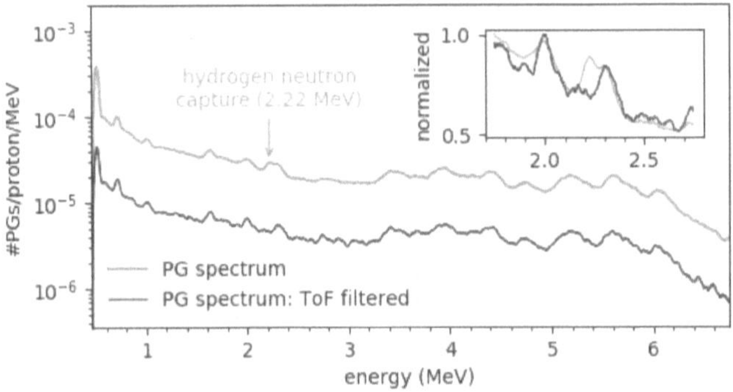

FIGURE 5.17: PG Spectrum with and without TOF background suppression. The yellow arrow serves as eye follow for the 2.22 MeV peak. Corresponding to the hydrogen neutron capture process, the 2.22 MeV peak is mitigated after the background suppression.

5.3.2.2 Anti-Coincidence Shield

The combined signal of four individual components from a total of eight BGOs was added and divided into two groups. A typical energy spectrum of one of the BGO groups (based on book data) is represented in figure 5.18. Only signals that arrived after the $CeBr_3$ event ($t_{BGO}^{event(i)} - t_{CeBr_3}^{event(i)} > 0$) were accepted. After calibration of BGOs with ^{137}Cs source, the electronic noise was analysed. Based on that analysis, an energy cut of ≈ 225 keV was applied. Thus, solely events above that cut were associated with escaped events from the primary detector.

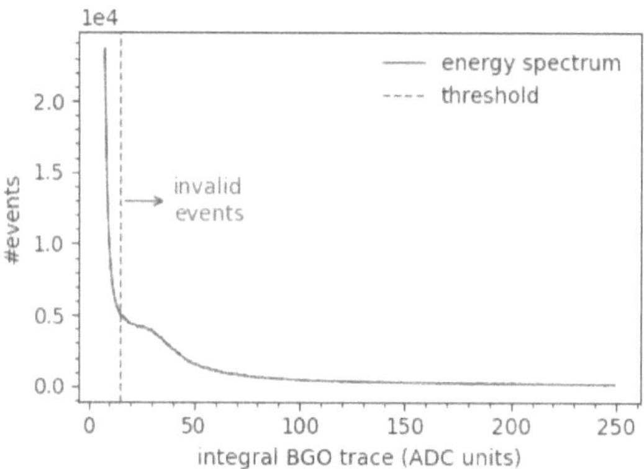

FIGURE 5.18: Energy spectrum of BGO events. The dashed line indicates the cut threshold adopted for invalid events.

The effect of the BGO background suppression on the original PG spectrum is illustrated in figure 5.19. The results show that active shielding is effective in reducing the background. Furthermore, the mitigation of SE and DE peaks are visible.

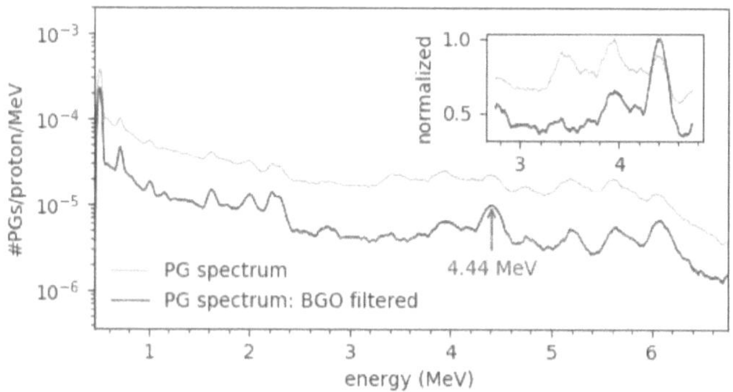

FIGURE 5.19: Energy spectrum of BGO events as function of energy.

Finally, figure 5.20 shows the summary of the background reduction methods. Discrete peaks can be observed in all stages of suppression due to the good energy resolution of *CeBr$_3$*.

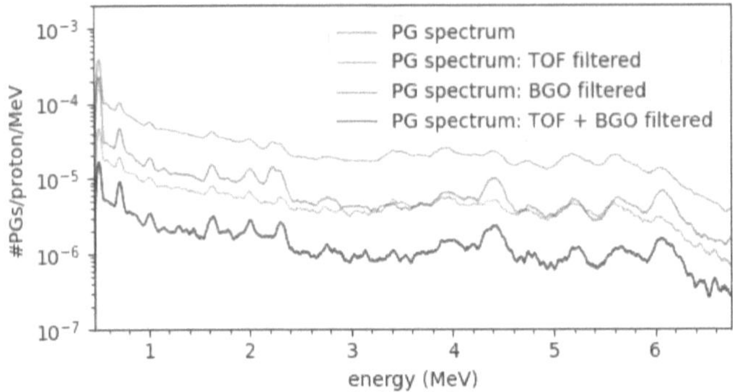

FIGURE 5.20: PG spectra with and without background suppression methods.

5.3.3 Peak Analysis

In this book, different algorithms to quantify the discrete peaks of PG spectra were used. To diminish the complexity of the gamma spectrum in peak analysis, background subtraction is essential. Thus, in this section, is briefly introduced. Additionally, the peak evaluations and identification methods used in this book are further dis-cussed.

5.3.3.1 Background Subtraction

PG spectra are drifted from zero mainly due to *Compton* scattering effect on target and detectors [102]. Thus, a technique of background subtraction (or baseline restoration) is essential. Techniques like sensitive non-linear iterative peak (SNIP) are usually effective in the baseline restoration. In this book, the background subtraction is achieved by the SNIP algorithm reported by P. Eilers and H. Boelens [103]. Before applying the SNIP algorithm it is required a pre-processing step, in which a transformation represented by equation 5.3 occurs, where i represents the channel, $y(i)$ are the energy counts in the channel, and $v(i)$ is the value after transformation [102, 103].

$$v(i) = \log\left[\log\left(\sqrt{y(i)+1}+1\right)+1\right] \tag{5.3}$$

The baseline is then evaluated in an iterative process that follows equation 5.4, where p is the *p-th* interaction, in which the transformed bin $v_{p-1}(i)$ is compared to the mean value given by the distance $\pm m$, where, m is a given parameter (usually the full width at half maximum) [102, 103].

$$v(i) = min \left\{ \frac{1}{2} \left[v_m(i - m) + v_{p-1}(i + m) \right], v_{p-1}(i) \right\} \tag{5.4}$$

An example of the SNIP application in a standard PG spectrum is illustrated in figure 5.21. The spectrum is shown for low energies because they are the region of interest in this book.

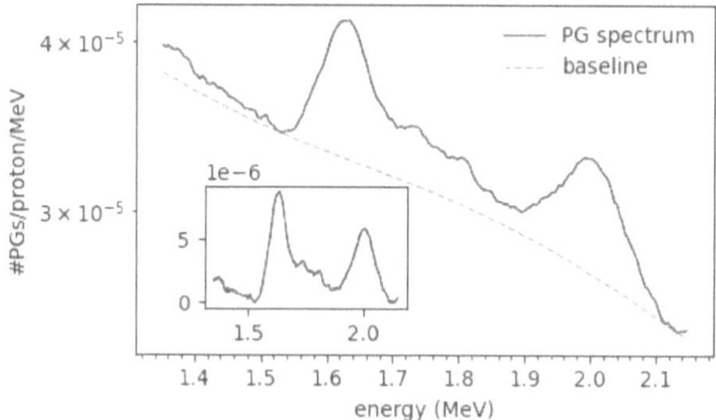

FIGURE 5.21: PG spectrum for low energies. Baseline calculated with a SNIP algorithm. The insert shows the PG spectrum after the baseline restoration.

5.3.3.2 Peak Evaluation

Evaluation of the PG spectrum peaks can by achieved by simple or more complex methods. In this book, the peak evaluation or quantification was performed with three different methods:

(a) total peak area

(b) peak fit in a region of interest

(c) multi peak fit

The total peak area method, (a), is carried out by a direct summation of the data points between prescribed limits. This method can be mathematically expressed by equation 5.5, where A is the peak area, C the number of counts in the i-th channel, n number of channels that belong to the peak region within the limits of channel L and U and m is the number of channels beyond each side of the peak region [16, 18].

$$A = \sum_{i=L}^{U} C_i - n \frac{\sum_{i=L-m}^{L-1} C_i + \sum_{i=U+1}^{U+m} C_i}{2m} \tag{5.5}$$

The peak fit in a region of interest, (b), is an analytic fit-based algorithm to the data points that comprises a single peak [18]. The fit algorithm used in this book was based on the non-linear least-squares method. The method throughout successive itera-tions try to minimize the differences between a function model, $F(x_i)$, and a given set of data points y_i (see equation 5.6). The *Gaussian* distribution was used to create the function model. Furthermore, the peak region was defined by the interval $[\mu - 2\sigma, \mu + 2\sigma]$, where μ and σ are the estimated mean and standard deviation of the *Gaussian* model. As a re-sult of method (b), the fit parameters, mean, sigma, height, the standard errors (when the uncertainties can be estimated) and χ^2 value are obtained.

$$\min \left(\sum |F(x_i) - y(i)|^2 \right) \tag{5.6}$$

In a similar way to the method (b), the multi-peak fit method, (c), uses the minimiza-tion function 5.6 to fit the data to more than one peak. In this case, the region of interest is broader, and it comprises several peaks. For this book, such region was defined by the interval [1.58 MeV, 1.84 MeV] in the PG spectrum. The goal is to distinguish the expected overlap between peaks. For results presented in chapter 6 and for both single peak fit and multi-peak fit it was used the parameter height rather than area as in method (a).

5.3.3.3 Peak Identification: F-statistics

The F-statistics or F-test compare nested models that fit the same d ataset. The nested model (the restricted model) is obtained from the full model by removing degrees of free-dom, df. It can be expressed mathematically by equation 5.7, where the subscript R and F denotes the restricted and full model respectively, RSS the residual sum of squares.

$$F_{df_R - df_F, df_F} = \frac{RSS_R - RSS_F}{RSS_F} \frac{df_F}{df_R - df_F} \tag{5.7}$$

In case of this book, the F-test was performed in the region from 1.58 MeV to 1.84 MeV, where the peaks 1.64 MeV and 1.78 MeV are expected. Thus, the full model was composed by three single models: a *Cauchy-Lorentz* distribution for each de-excitation peak with parameters amplitude, center, and sigma, and a fixed constant model with a parameter value of zero. The goal is to know if the data can be explained without the 1.78 MeV peak. Therefore, the restricted model was obtained from the full model by the elimination of the distribution parameters associated with 1.78 MeV. The null and alternative hypothesis are:

- *H0*: parameters for the omitted peak are all zero;

- *H1*: at least one omitted parameter is not zero.

Chapter 6

Results and Discussion

Results and discussion are presented in this chapter in a way where their discussion follows the results. All experimental data were taken in HIT as a direct collaboration between DKFZ and HIT.

6.1 Preliminary Analysis of ^{28}Si Prompt Gamma-Ray Lines

6.1.1 Direct Beam

As the first approach, several samples of H_2O + SiO_2 based mixtures were prepared. An optimal mixture regarding its saturation and suitability for rectal balloon application was achieved. As mention in chapter 5 the mixture was in a ration of 3 of water to 2 of silicon dioxide. In a preliminary analysis the balloon filled with H_2O + SiO_2 was directly irradiated with a proton beam energy and intensity of $E_p = 48.12\,\mathrm{MeV/u}$ and $I_p = 2.0 \times 10^8$ protons/s, respectively.

FIGURE 6.1: Experimental setup of the rectal balloon directly irradiated. The right detector is the main detector and the left detector the detector that serves as validation.

Figure 6.1 shows a photo of the experimental setup. The target and all instrumentation were aligned to the beam isocenter. The distance between the detectors and beam isocenter was $d_{detector} = 15 \pm 0.05\,cm$. The extra detector served as validation. Throughout the book, only results regarding the one detector (see figure 6.1, right detector, and chapter 5) are presented. In this sub-section, both detectors PG spectra were computed. Their results are depicted in figure 6.2. Apart from activation peak, the 1.78 MeV peak from de-excitation of $^{28}Si^*$ stands out for lower energies. It can also be noticed that both detectors have a similar response. In the lower energy region ($< 2\,MeV$) of the PG spectrum, several lines induced by protons in ^{28}Si are visible and affect the PG spectrum. However, they overlap with carbon/oxygen lines and SE/DE of high energetic peak lines.

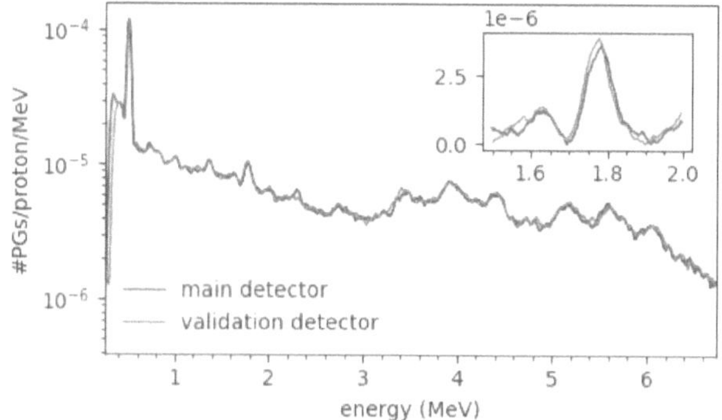

FIGURE 6.2: PG Spectrum of rectal balloon filled with H_2O + SiO_2 and irradiated with a proton beam of $E_p = 48.12\,MeV/u$ and $I_p = 2.0 \times 10^8\,protons/s$.

6.1.2   Flask of H_2O vs Flask of H_2O + SiO_2 vs Balloon H_2O + SiO_2

As expected the 1.78 MeV ^{28}Si peak is the best candidate for the purposes of this book. To evaluate the effect of the tissue, it was added two flasks filled with H_2O between the rectal balloon and the beam nozzle. A photo of the experimental setup is presented in figure 6.3. In order to distinguish the 1.78 MeV ^{28}Si peak, the balloon was replaced by a flask filled with H_2O and a validation flask filled with H_2O + SiO_2 mixture (see figure 6.3). As in the previous sub-section, the targets and the detectors at $d_{detector} = 15 \pm 0.05\,cm$ were aligned with isocenter. The targets were irradiated with proton beam of $E_p = 113.58\,MeV/u$ and $I_p = 2.0 \times 10^8\,protons/s$.

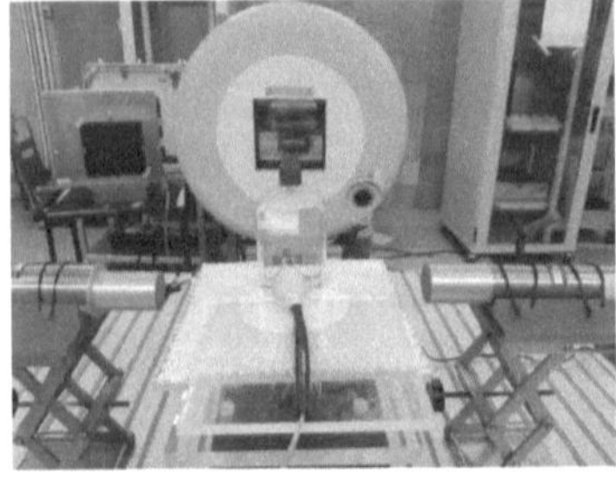

(A) rectal balloon (H_2O + SiO_2)

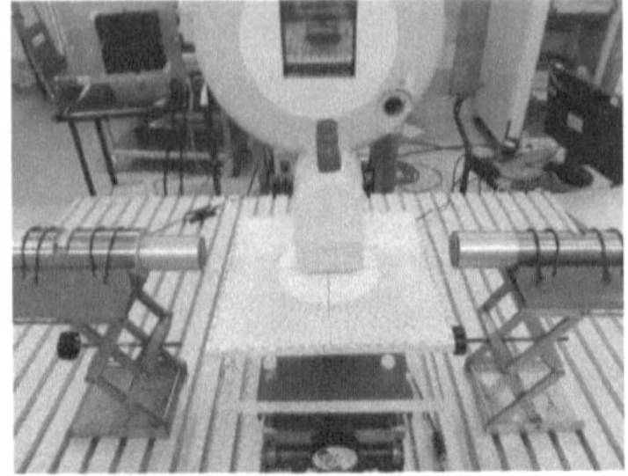

(B) flask (H_2O or H_2O + SiO_2)

FIGURE 6.3: Experimental setup of two flasks filled with H_2O followed either by a rectal balloon or a third flask. The balloon and the third flask were filled with H_2O + SiO_2 mixture.

The first analysis falls on the two extra flasks effect study. The PG spectra for the rectal balloon with and without the two H_2O flasks are reproduced in figure 6.4. Regarding the use of the two flasks, the most significant differences are higher PG production (due to the larger beam path length in the target); increase and the emergence of new $^{16}O/^{12}C$ discrete lines (e.g., 1.64 MeV peak from de-excitation of $_{14}N^*_{2.31}$, see appendix A); consequently *Compton* continuum increase; and the ^{28}Si discrete line are smeared out due to background.

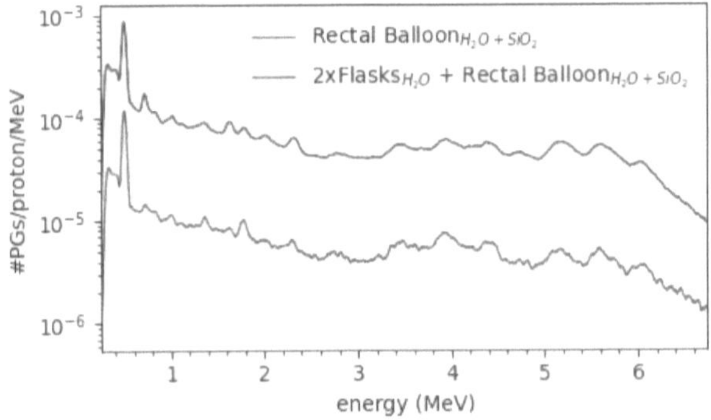

FIGURE 6.4: PG spectra of the rectal balloon filled with H_2O + SiO_2 irradiate with and without two H_2O flasks between the beam and the balloon. Without the flask, $E_p = 48.12$ MeV/u and $I_p = 2.0 \times 10^8$ protons/s, with flasks, $E_p = 113.58$ MeV/u and $I_p = 2.0 \times 10^8$ protons/s.

When comparing the two flasks plus rectal balloon or third H_2O flask setup, one can expect a similar amount of PG production. Such correlation can be seen in figure 6.5. The zoom-in inset was obtained after SNIP-based background subtraction. Analysing the inset, it is apparent that background subtraction is achieved without significantly affecting the peaks features. The 1.64 MeV peak magnitude is the same for both. For high energies (> 2 MeV) the PG spectra of 6.5 are similar. However, for lower energies, some differences are observed. These differences can be explained by the PG continuum associated with the ^{28}Si lines.

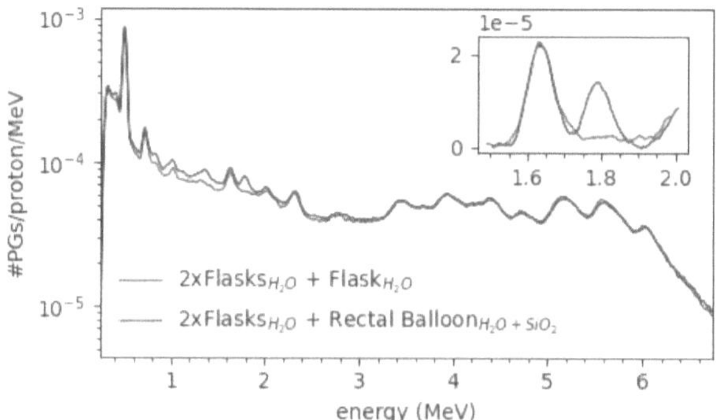

FIGURE 6.5: PG spectra from the experimental setup shown on figure 6.3, where the balloon was replaced by an extra flask filled with H_2O (not shown). Proton beam irradiation with $E_p = 113.58$ MeV/u and $I_p = 2.0 \times 10^8$ protons/s.

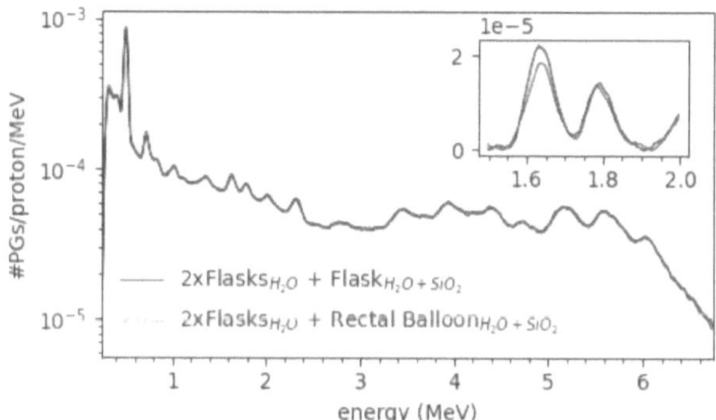

FIGURE 6.6: PG spectra from the experimental setup shown on figure 6.3 (b). The extra flask was filled either by H_2O or $H_2O + SiO_2$. Proton beam irradiation with $E_p = 113.58$ MeV/u and $I_p = 2.0 \times 10^8$ protons/s.

Using the same experimental setup (two flasks plus target), two PG spectra were obtained for the flask and the rectal balloon filled with $H_2O + SiO_2$ mixture. The results are shown in figure 6.6. The two PG spectra overlap for all energy range showing that different encapsulations of the mixture do not influence the PG spectrum. Results confirm the reproducibility of PG spectrums and are a validation for the next section, in which different samples within flask or rectal balloon are compared.

6.2   Samples Analysis

Notwithstanding the good results achieved by the $H_2O + SiO_2$ (3 of H_2O to 2 of SiO_2) optimal mixture, new samples were prepared: $H_2O + SiO_2$ mixture in a ratio of 2 of H_2O to 1 of SiO_2 and in a ratio of $\approx$ 1 of H_2O to 1 of SiO_2, and a sample of commercial silicone sealant. They were directly irradiated with a proton beam energy and intensity of 48.12 MeV/u and 2.0×10^8 protons/s, respectively. The measured PG spectra are compared against the previous optimal mixture measurements. The results are depicted in figure 6.7. As the ratio of $H_2O + SiO_2$ increases, so does the 1.78 MeV ^{28}Si peak magnitude. The commercial silicon and the $H_2O + SiO_2$ mixture in a ratio of $\approx$ 1 of H_2O to 1 of SiO_2 were a benchmark, since their repetitive insertion in the balloon was not feasible. When compared with the mixtures, the commercial silicone sealant has lower oxygen concentration. This observation translates in the reduced 6.13 MeV ^{16}O peak magnitude (and correspondent SE and DE) in PG spectrum (energy range $\approx$ 5 to 6.5 MeV).

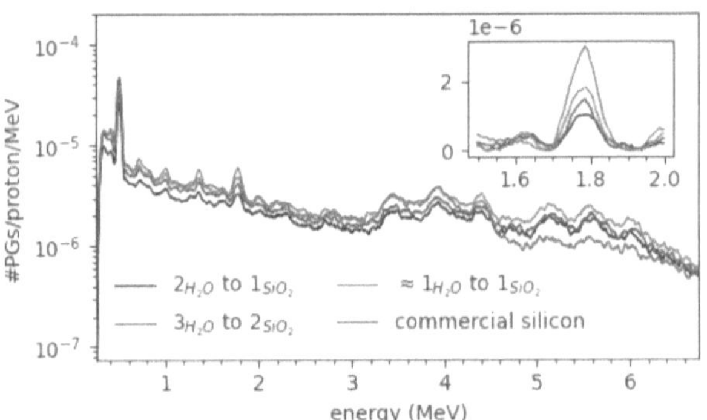

FIGURE 6.7: PG Spectrum for different samples. All samples were irradiated with a proton beam of $E_p = 48.12$ MeV/u and $I_p = 2.0 \times 10^8$ protons/s.

6.3   In-Vitro Single Spot Beam Binary Analysis

The previous studies showed that the rectal balloon filled with H_2O + SiO_2 mixture
is suitable for in-vitro experiments. Subsequent studies involve the use of a prostate-like
phantom, CIRS070L (see chapter 5), to evaluate how the detection of 1.78 MeV ^{28}Si peak
can be translated into a better clinical scenario in proton treatment for prostate cancer.
For this very purpose, the experimental setup in this section was prepared to be irradiated by bilateral/anterior and single spot beams. The anterior and bilateral beams are
represented in figure 6.8 by blue and yellow arrows, respectively. Six beams were used
in total. Dashed lines in the figure represent prostate irradiations and solid lines rectal
balloon irradiations.

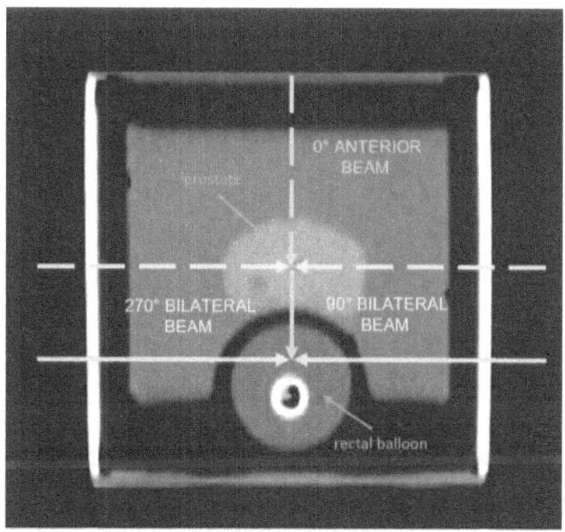

FIGURE 6.8: Axial view of the prostate training phantom (*CIRS Model 070L*) CT image.
The beams are represented either by solid or dashed lines in a total of 6 beams. Solid and
dashed lines mean that *Bragg* peaks stop in the rectal balloon or prostate, respectively.

6.3.1   Bilateral Beams: 90°and 270°

A photo of the experimental setup regarding BLs is displayed in figure 6.9 for both
90°and 270°angles. The 14 spills irradiation of the prostate and rectal balloon was performed by a proton beam of $E_p = 93.02$ MeV/u, and $I_p = 2.0 \times 10^8$ protons/s. The different single spots for prostate and rectal balloon in the BL setup were achieved by a beam
translation of 3 cm.

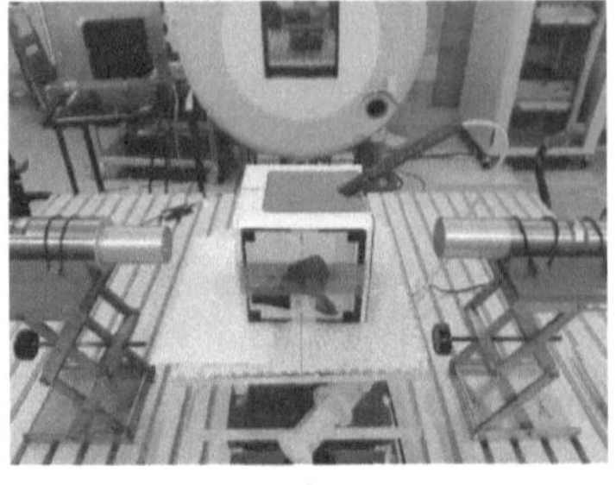
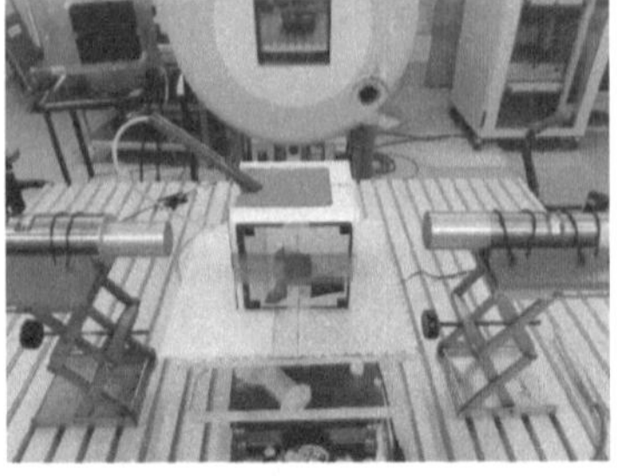

(A) 90°direction   (B) 270°direction

FIGURE 6.9: Experimental setup used for bilateral beam irradiation in a patient-like prostate phantom.

In the PGS analysis of the configuration presented in figure 6.9 both, main (at right) and validation (at left) detector were used. The PG spectra for main and validation detectors are presented in figure 6.10 and 6.11 for 90°and 270°direction, respectively.

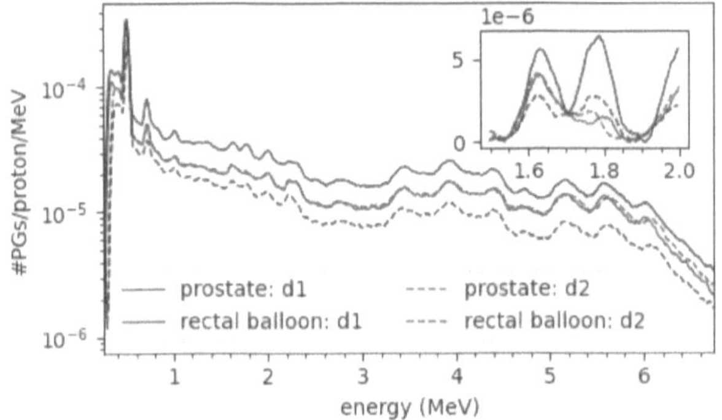

FIGURE 6.10: PG spectra of prostate and rectal balloon with bilateral beam direction of 90°. The legends d1 and d2 stands for main and validation detector, respectively. The patient-like prostate phantom was irradiated with a proton beam of $E_p = 93.02\,\text{MeV}/\text{u}$ (balloon) and $I_p = 2.0 \times 10^8\,\text{protons}/\text{s}$.

One can observe different prominences in the 1.78 MeV ^{28}Si peak for different detectors. The detector closer to the balloon (main detector for the 90° direction and validation detector for the 270° direction) has a higher prominence. This suggests that detectors geometric position has a significant impact on the outcome of ^{28}Si peak magnitude. Notwithstanding with geometric restrain, the ^{28}Si is well visible at least with one of the detectors.

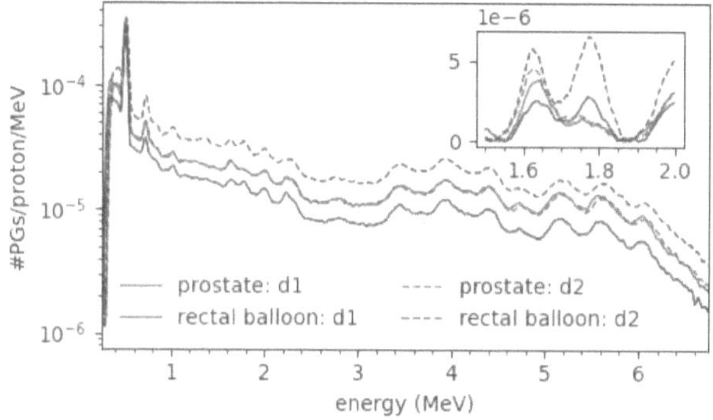

FIGURE 6.11: PG Spectrum of prostate and rectal balloon irradiation with bilateral beam direction of 270°. The legends d1 and d2 stands for main and validation detector, respectively. The patient-like prostate phantom was irradiated with a proton beam of $E_p = 93.02\,\text{MeV/u}$ (balloon) and $I_p = 2.0 \times 10^8$ protons/s.

6.3.2 Anterior Beams: 0°

A photo of the experimental setup regarding anterior beams is displayed in figure 6.12. Although the beam trigger is visible in figure such device was not used in this subsection analysis. The 14 spills irradiation of the prostate and rectal balloon was performed by a proton beam of $E_p = 98.27\,\text{MeV/u}$ (prostate) and $E_p = 112.25\,\text{MeV/u}$ (balloon), and $I_p = 1.8 \times 10^7$ protons/s.

FIGURE 6.12: Experimental setup used for anterior beam irradiation in a patient-like prostate phantom.

The PG spectra for prostate and rectal balloon irradiation with anterior beams are depicted in figure 6.13. The 1.78 MeV ^{28}Si is easily discriminated despite the overlap with the 1.64 MeV peak. One can also observe a different ^{28}Si peak shape. *Compton* scattering is the main reason for the observed irregularities. The differences in spectra background can be explained by the different energies (ranges) used.

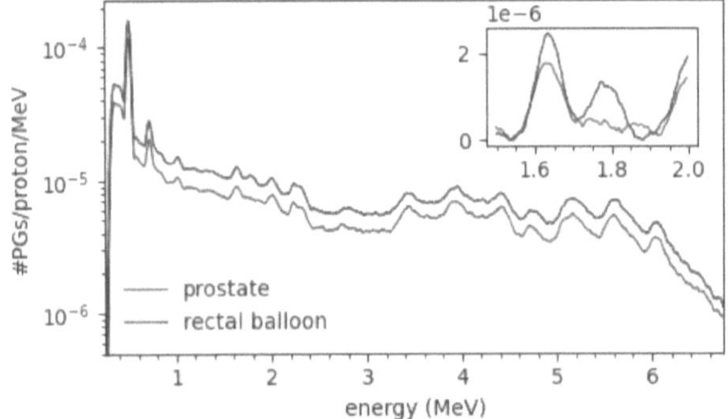

FIGURE 6.13: PG spectra of prostate and rectal balloon irradiation with anterior beams.

The results presented in this section show that 1.78 MeV ^{28}Si peak can be exploited to infer about the position of the beam, or more specifically if the *Bragg* peak reaches or not the balloon. The *Si* peak evaluation is performed in the next section using a similar setup to the one used in this sub-section.

6.4   1.78 MeV ^{28}Si Prompt Gamma-Ray Line Evaluation

In the previous section, single spot irradiations for two different beam setups, bilateral and anterior beams, were studied. The results show that 1.78 MeV ^{28}Si peak prominence is rather high for the bilateral setup than for anterior. This indicates that ^{28}Si peak magnitude can be quantified in order to know how deep is the *Bragg* peak within the rectal balloon. In this section, a preliminary quantification of the ^{28}Si peak is performed.

6.4.1   1.78 MeV ^{28}Si Peak Presence and Magnitude vs Depth

In order to evaluate the magnitude of the 1.78 MeV ^{28}Si peak, several PG spectra from different proton energies were taken. The anterior beam setup with an angle of 0°was used. Additionally, both beam trigger and anti-coincidence shield were used as an option for background suppression. The experimental setup is shown in figure 6.14.

FIGURE 6.14: Experimental setup used to obtain different PG spectra for the incidence of anterior proton beam with different energies in a patient-like prostate phantom. Both, beam trigger and anti-coincidence shield were used as background suppression.

A main campaign of measurements was firstly performed, followed by a validation campaign. Both campaigns used the rectal balloon filled with $H_2O + SiO_2$ mixture. Additionally, the validation campaign was completed by a reference measurement using the rectal balloon filled with just H_2O. The beam with 14 spills had an intensity of 8.0×10^7 which totals a number of protons of $N_p \approx 5.6 \times 10^9$. The different proton energies used in both campaigns are listed in table 6.1. In the table is also presented the *Bragg* peak position relative to the distal edge (rectal wall) of the *CIRS Model 070L* phantom. Furthermore, the normalized dose distribution for two of the energies (94.54 MeV and 112.25 MeV) in table 6.1 is illustrated in figure 6.15.

The data of several PG spectra were first analysed in a qualitative way with and without background suppression. In figure 6.16 is presented the PG spectra for proton energies of 101.18, 105.43, 108.88, 112.25 and 115.55 MeV/u. The same spectra are presented without and with all background suppression techniques. Moreover, a region of interest around the 1.78 MeV ^{28}Si was defined for the spectra presentation. Notwithstanding the effectiveness of the filters, due to low statistics, the PG spectrums get distorted. When background suppression with TOF, BGO or TOF + BGO the number of protons are reduced approximately to 13%, 32% and 6%, respectively, of the original number of protons. Low statistics bring undesired effects in the quantification of the peak. Therefore, only background suppression based on the anti-coincident shield is an option.

TABLE 6.1: Table with beam energies used in the study of peak magnitude vs depth (main and validation campaign). The relative *Bragg* peak distance to the distal edge of the *CIRS Model 070L* phantom and the target structure associated is also presented.

energy (MeV/u)	main campaign	validation campaign	relative Bragg peak distance (cm)	Bragg peak position (structure)
86.72	✓		-7.4	prostate
90.70	✓		-6.9	prostate
94.54	✓		-6.4	prostate
96.05	✓		-6.2	prostate
97.53	✓		-6.0	prostate
98.27		✓	-5.9	prostate
99.01	✓		-5.8	prostate
100.46	✓		-5.6	prostate
101.18	✓	✓	-5.4	rectal wall
103.32	✓	✓	-5.2	rectal wall
104.03	✓		-5.1	rectal wall
104.73	✓	✓	-5.0	rectal wall
105.43	✓	✓	-4.9	ballooon
106.12	✓	✓	-4.8	ballooon
107.51	✓	✓	-4.6	ballooon
108.88	✓	✓	-4.4	ballooon
112.25	✓	✓	-3.9	ballooon
115.55	✓	✓	-3.4	ballooon
118.78	✓		-2.9	handle
121.95	✓		-2.4	handle
125.06	✓		-1.9	ballooon
128.11	✓		-1.4	ballooon
131.11	✓		-0.9	rectal wall
134.06	✓		-0.4	rectal wall

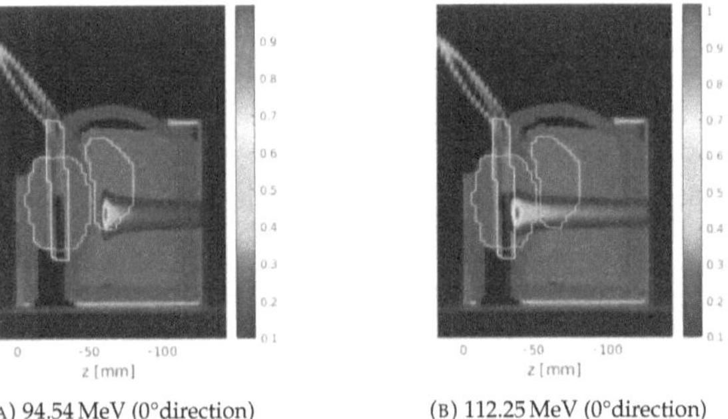

(A) 94.54 MeV (0°direction)   (B) 112.25 MeV (0°direction)

FIGURE 6.15: Normalized dose distribution for a proton beam of 94.54 MeV and 112.25 MeV. The structures in the prostate phantom are identified with orange (prostate), green (balloon) and yellow contour lines. Data obtained from Monte Carlo simulations.

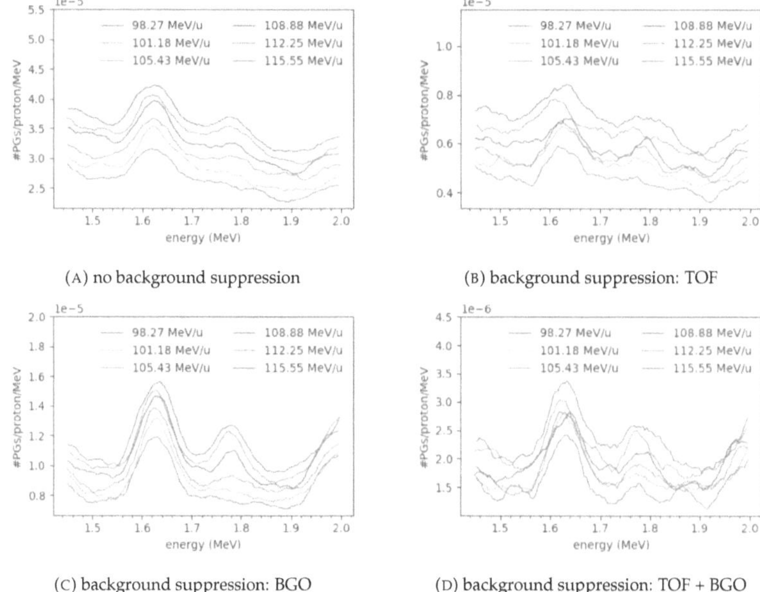

(A) no background suppression

(B) background suppression: TOF

(C) background suppression: BGO

(D) background suppression: TOF + BGO

FIGURE 6.16: PG spectrum for a proton beam with different energies. The spectrum is zoomed in a region of interest related with 1.78 MeV ^{28}Si peak emission. Different combinations of background suppression was used for a better insight of its effect in the spectrum.

The 1.78 MeV ^{28}Si peak presence was evaluated with the F-test with a confidence level of 99.99 % ($> 3\sigma$). The results for different energies/ranges and with/without background suppression based on anti-coincidence shield are depicted in figures 6.17 and 6.18. The F-test with some exceptions allows checking the presence of the *Bragg* peak after ≈ 4 mm within the rectal balloon. In the presence of the handler, the statistics variations are high, and the results within that region cannot be considered significant for the analysis carried out in this book. The results with background suppression have more statistical significance regarding the ^{28}Si peak presence. On the other hand, when looking to the 6.17, in the reference measurements, a type I error is more likely to happen.

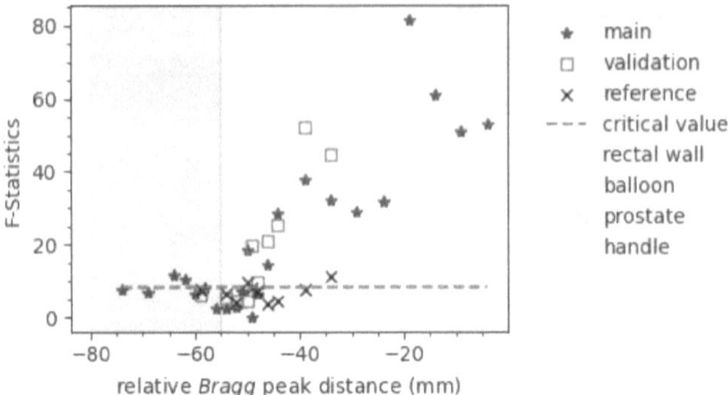

FIGURE 6.17: F-test vs relative *Bragg* peak distance for measurements without background suppression. The critical value was set with a confidence level of 99.99 %.

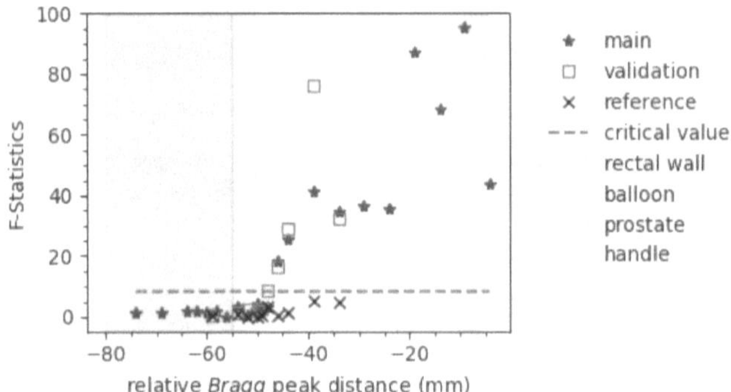

FIGURE 6.18: F-test vs relative *Bragg* peak distance for measurements with background suppression (BGO-based). The critical value was set with a confidence level of 99.99 %.

To quantify the 1.78 MeV ^{28}Si peak for different energies/ranges the methods (a), (b), (c) mention in chapter 5 were used. The results with and without background suppression based on anti-coincidence shield are depicted in figures 6.19 and 6.20, respectively. The behavior of the points in the plots shows great potential for predicting single-spot proton range in prostate cancer, or more specifically, within the balloon. The handle effect on data is visible in all cases. The mean relative error between the main and validation campaign was calculated for both cases with and without background suppression. Its values were 11 % and 2 %, respectively. The relative error for the case of no background suppression indicates a good reproducibility between results.

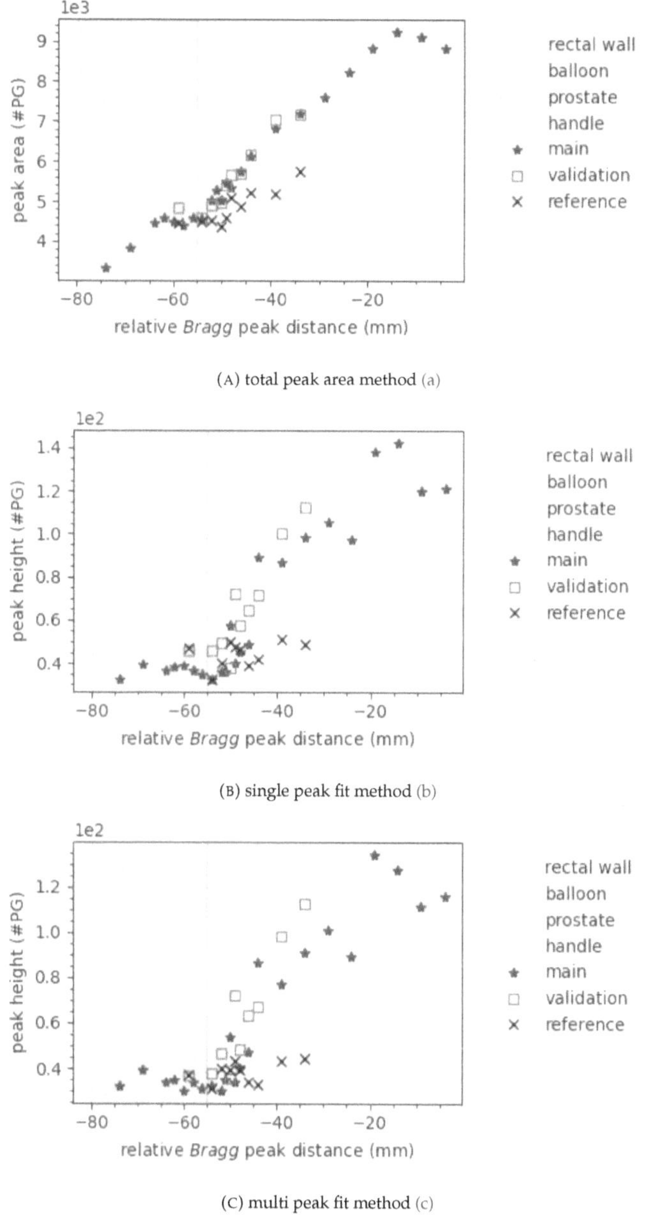

(A) total peak area method (a)

(B) single peak fit method (b)

(C) multi peak fit method (c)

FIGURE 6.19: Peak area and height vs relative *Bragg* peak distance for measurements without background suppression.

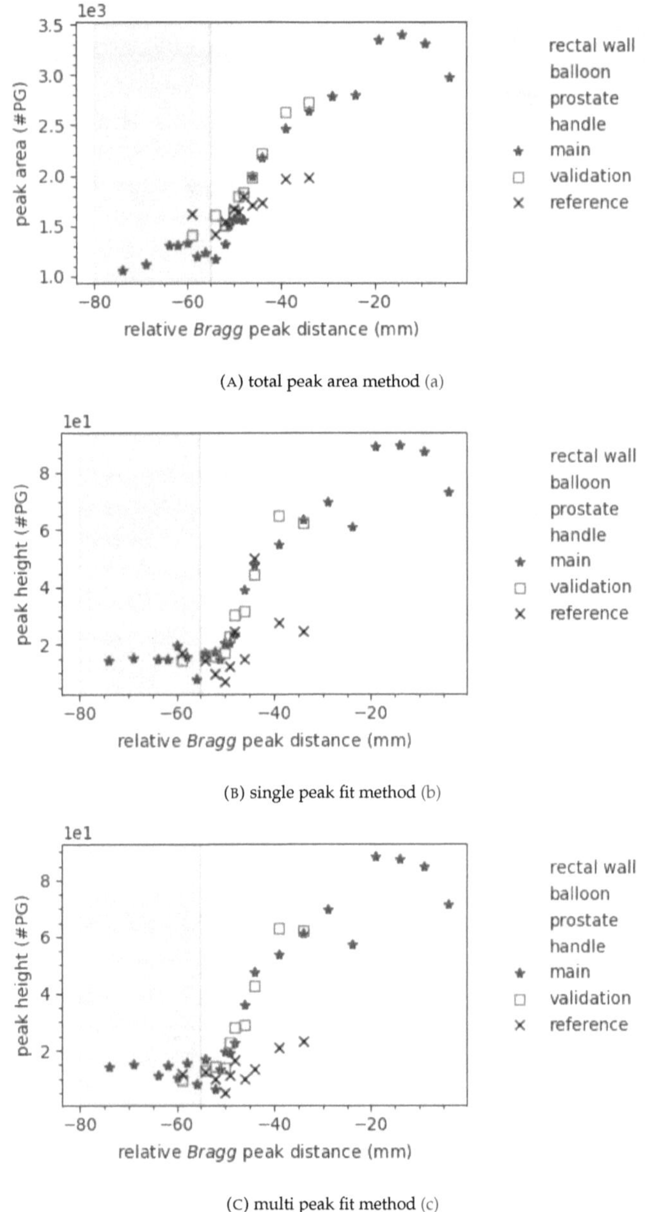

(A) total peak area method (a)

(B) single peak fit method (b)

(C) multi peak fit method (c)

FIGURE 6.20: Peak area and height vs relative *Bragg* peak distance for measurements with background suppression (BGO-based).

A polynomial model of degree 2 was created (for both cases, with and without background suppression) based on campaigns measurements within the region of $-46\,$mm to $-34\,$mm. Then, both models were compared with the expected results (see table 6.2). The result of the fit is shown in figure 6.21.

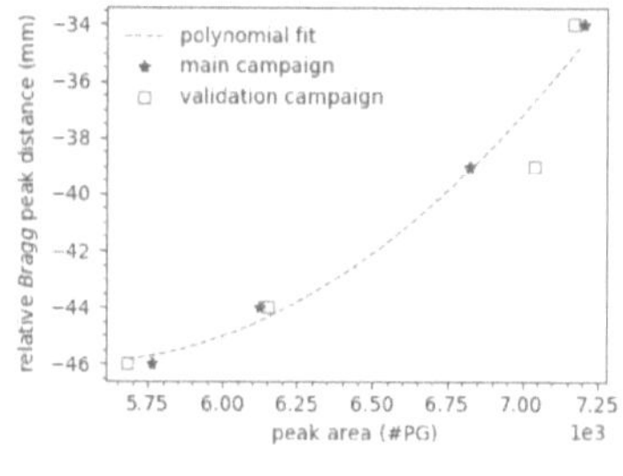

(A) no background suppression

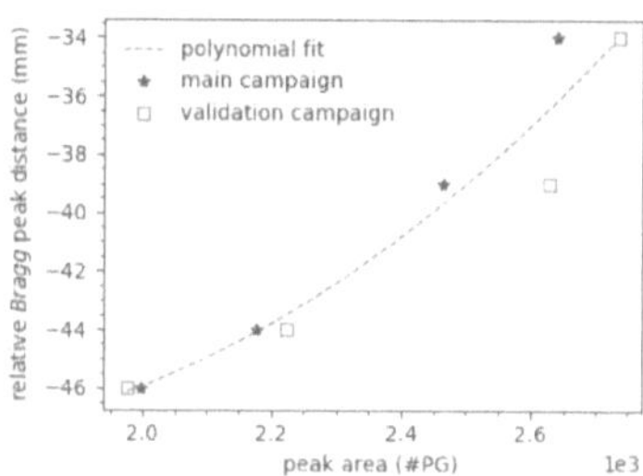

(B) background suppression: BGO

FIGURE 6.21: Polynomial fit of second order to the main and validation campaign measurements of figure 6.19 (A) and 6.20 (A), and within the region of $-46\,$mm to $-34\,$mm

TABLE 6.2: Results of the polynomial fit model of campaigns measurements when applied to the main and validation campaign.

expected distances (mm)	distances without background suppression		distances with background suppression	
	main campaign (mm)	*validation campaign (mm)*	*main campaign (mm)*	*validation campaign (mm)*
-46.0	-45.7	-45.8	-46.0	-46.1
-44.0	-44.5	-44.4	-44.1	-43.5
-39.0	-39.2	-36.7	-39.7	-36.4
-34.0	-34.6	-35.1	-36.1	-34.0

6.5 Final Discussion

The studies presented in this chapter demonstrated the feasibility of range-monitoring based on PGS and 1.78 MeV ^{28}Si magnitude. Qualitatively the ^{28}Si peak was observed for every scenario. In-vitro measurements with the prostate-like phantom shown the importance of a good geometric position of the detectors.

The quantitative analysis focused on the peak presence and magnitude. The peak presence was estimated with a level of confidence of $> 3\sigma$. Due to the test's high accuracy, negligible false positives were observed in a reference sample with only H_2O. Nevertheless, the peak presence was detected for a depth of $< 4mm$ within the balloon.

The magnitude analysis revealed the potential of using it to infer about the particle range within the rectal balloon. Fitting a polynomial model to the data allows the estimation of the range within the balloon. The root mean square error (RMSE) was calculated for both situations, with and without background suppression. Without background suppression, the RMSE was 0.44 mm and 1.3 mm for the main and validation campaign, respectively. With background suppression, the RMSE was 1.1 mm and 1.3 mm for the main and validation campaign, respectively.

Although more studies are needed, the studies hereby presented unveil the potential for predicting single-spot proton range within the balloon, and therefore *in-vivo* range monitoring.

The possibility of *in-vivo* range monitoring in patients on a daily basis will allow the clinician to reduce tumor margins, which subsequently reduces radiation dose given to healthy surrounding organs. The main advantage of smaller tumor margins is a reduction in the complications due to lower doses given to healthy organs. Furthermore, a reliable system will allow correction of range errors between fractions with re-planning enhancing the quality of treatment.

Chapter 7

Conclusion and Future Work

It was presented the first study of an *in-vitro* setup entailed with proton beams, PGS measurements, and the use of a rectal balloon filled with a ^{28}Si-based solution within a prostate-like phantom. The presence of 1.78 MeV of ^{28}Si PG line on the energy spectra after a qualitative analysis was quantified for different scenarios.

As the first approach, the qualitative analysis was focused on simple setups where ^{28}Si-based samples were either directly irradiated or irradiated through a certain thickness of tissue-equivalent material. The peak was visualized in all cases. Furthermore, the used mixture of 3 of H_2O to 2 of SiO_2 gave out good results regarding peak prominence. An initial study regarding a prostate-like phantom followed. The study revealed the strong dependence of the geometric detector position. Nevertheless, the peak was observed for anterior and bilateral beams. The quantitative analysis of the 1.78 MeV ^{28}Si peak shown great potential for predicting single-spot proton range within the balloon.

The 1.78 MeV ^{28}Si peak online measurement can provide a way of *in-vivo* range monitoring. However, improvements can be achieved with more research on it. The PGS system needs to be evaluated in different clinical scenarios of prostate cancer treated by proton and other ions. This involves studying several parameters, such as evaluate best beam angles for the use of PGS in prostate cancer, evaluate how to maximize PGS signal as a function of the tumor location and depth, perform tumor margin assessment, to understand what is smallest acceptable margins for surrounding organs. Additionally, it is important to weigh the advantages/disadvantages caused by the rectal balloon filled with ^{28}Si-based sample. Finally, one of the most important studies to be clinically applicable is the modulation and characterization of the PGS system using Monte Carlo simulations.

Appendix A

Proton Induced Gamma-Ray Lines in ^{12}C, ^{14}N, ^{16}O and ^{28}Si

TABLE A.1: Nuclear proton induced de-excitation gamma-ray lines in ^{12}C (only the most relevant to the book were selected) [13, 23–25]

Target: ^{12}C	
nuclear reaction de-excitations	emission (MeV)
$^{10}\text{B}^{*}_{0.718} \longrightarrow {}^{10}\text{B}_{\text{g.s.}}$	0.718
$^{10}\text{B}^{*}_{1.74} \longrightarrow {}^{10}\text{B}^{*}_{0.718}$	1.02
$^{15}\text{O}^{*}_{7.56} \longrightarrow {}^{15}\text{O}^{*}_{6.18}$	1.38
$^{14}\text{N}^{*}_{3.95} \longrightarrow {}^{14}\text{N}^{*}_{2.31}$	1.64
$^{15}\text{N}^{*}_{7.16} \longrightarrow {}^{15}\text{N}^{*}_{5.27}$	1.88
$^{11}\text{C}^{*}_{2.00} \longrightarrow {}^{11}\text{C}_{\text{g.s}}$	2.00
$^{12}\text{C}^{*}_{4.44} \longrightarrow {}^{12}\text{C}_{\text{g.s}}$	4.44
$^{15}\text{N}^{*}_{5.27} \longrightarrow {}^{15}\text{N}_{\text{g.s}}$	5.27

TABLE A.2: Nuclear proton induced de-excitation gamma-ray lines in ^{14}N (only the most relevant to the book were selected) [23–25].

Target: ^{14}N	
nuclear reaction de-excitations	emission (MeV)
$^{15}O^{*}_{7.56} \longrightarrow {}^{15}O^{*}_{6.18}$	1.38
$^{14}N^{*}_{3.95} \longrightarrow {}^{14}N^{*}_{2.31}$	1.64
$^{14}N^{*}_{9.15} \longrightarrow {}^{14}N^{*}_{7.15}$	2.00
$^{14}N^{*}_{2.31} \longrightarrow {}^{14}N_{g.s}$	2.31
$^{12}C^{*}_{4.44} \longrightarrow {}^{12}C_{g.s}$	4.44

TABLE A.3: Nuclear proton induced de-excitation gamma-ray lines in ^{16}O (only the most relevant to the book were selected) [13, 23–25].

Target: ^{16}O	
nuclear reaction de-excitations	emission (MeV)
$^{10}B^{*}_{0.718} \longrightarrow {}^{10}B_{g.s.}$	0.718
$^{10}B^{*}_{1.74} \longrightarrow {}^{10}B^{*}_{0.718}$	1.02
$^{15}O^{*}_{7.56} \longrightarrow {}^{15}O^{*}_{6.18}$	1.38
$^{14}N^{*}_{3.95} \longrightarrow {}^{14}N^{*}_{2.31}$	1.64
$^{15}N^{*}_{7.16} \longrightarrow {}^{15}N^{*}_{5.27}$	1.88
$^{14}N^{*}_{2.31} \longrightarrow {}^{14}N_{g.s}$	2.31
$^{16}O^{*}_{8.87} \longrightarrow {}^{16}O^{*}_{6.13}$	2.74
$^{12}C^{*}_{4.44} \longrightarrow {}^{12}C_{g.s}$	4.44
$^{15}N^{*}_{5.27} \longrightarrow {}^{15}N_{g.s}$	5.27
$^{16}O^{*}_{6.13} \longrightarrow {}^{16}O_{g.s}$	6.13

TABLE A.4: Nuclear proton induced de-excitation gamma-ray lines in ^{28}Si (only the most relevant to the book were selected) [23–26, 104].

Target: ^{28}Si	
nuclear reaction de-excitations	emission (MeV)
$^{27}\mathrm{Si}^*_{0.7803} \longrightarrow {}^{27}\mathrm{Si}_{\mathrm{g.s.}}$	0.780
$^{27}\mathrm{Al}^*_{0.844} \longrightarrow {}^{27}\mathrm{Al}^*_{\mathrm{g.s.}}$	0.844
$^{27}\mathrm{Si}^*_{0.957} \longrightarrow {}^{27}\mathrm{Si}_{\mathrm{g.s.}}$	0.957
$^{27}\mathrm{Al}^*_{1.01} \longrightarrow {}^{27}\mathrm{Al}_{\mathrm{g.s.}}$	1.01
$^{31}\mathrm{P}^*_{1.27} \longrightarrow {}^{31}\mathrm{P}_{\mathrm{g.s.}}$	1.27
$^{24}\mathrm{Mg}^*_{1.37} \longrightarrow {}^{24}\mathrm{Mg}_{\mathrm{g.s.}}$	1.37
$^{28}\mathrm{Si}^*_{1.78} \longrightarrow {}^{28}\mathrm{Si}_{\mathrm{g.s}}$	1.78
$^{27}\mathrm{Al}^*_{2.21} \longrightarrow {}^{27}\mathrm{Al}_{2.21}$	2.21
$^{20}\mathrm{Ne}^*_{4.25} \longrightarrow {}^{20}\mathrm{Ne}^*_{1.63}$	2.61
$^{28}\mathrm{Si}^*_{6.88} \longrightarrow {}^{28}\mathrm{Si}^*_{1.78}$	5.01
$^{28}\mathrm{Si}^*_{6.88} \longrightarrow {}^{28}\mathrm{Si}_{\mathrm{g.s}}$	6.88